Computational Molecular Science and Engineering Forum 2016

Core Programming Area at the 2016 AIChE Annual Meeting

San Francisco, California, USA
13 - 18 November 2016

ISBN: 978-1-5108-3431-6

Printed from e-media with permission by:

Curran Associates, Inc.
57 Morehouse Lane
Red Hook, NY 12571

CURRAN ASSOCIATES INC. proceedings.com

Printed by Curran Associates, Inc. (2017)

For permission requests, please contact AIChE
at the address below.

AIChE
120 Wall Street, FL 23
New York, NY 10005-4020

Phone: (800) 242-4363
Fax: (203) 775-5177

www.aiche.org

Additional copies of this publication are available from:

Curran Associates, Inc.
57 Morehouse Lane
Red Hook, NY 12571 USA
Phone: 845-758-0400
Fax: 845-758-2633
Email: curran@proceedings.com
Web: www.proceedings.com

Computational Molecular Science and Engineering Forum 2016

Core Programming Area at the 2016 AIChE Annual Meeting

San Francisco, California, USA
13 - 18 November 2016

AIChE
The Global Home of Chemical Engineers

TABLE OF CONTENTS

461800 Accelerating Materials Design: Computer Simulations and QSAR Modeling

Abstract:

Research Interests:

My future research group will focus on designing soft materials and processes that can help us face critical health, energy and environment challenges. Computer simulations and QSAR modeling have shown their ability to accelerate materials discovery and process design by revealing critical molecular-level mechanisms, providing rational design principles, and identifying key molecular architectures. My three research projects are to design (1) zwitterionic electrolytes for Li^+batteries and solar cells, (2) peptide materials for high-performance medical diagnostics and (3) amyloid-forming peptides for industrial catalysts using computer simulations and QSAR modeling.

My PhD and postdoctoral research experiences equip me with the expertise to apply/develop computer simulations/QSAR modeling in cooperation with experimentalists. My PhD research in Prof. Shaoyi Jiangâ€™s group at the University of Washington focused on understanding and designing zwitterionic anti-biofouling materials using computer simulations and experiment. Zwitterionic materials possess cationic and anionic groups in one moiety. We developed force field parameters for zwitterionic moieties, investigated hydration, protein interaction, ion interaction, and self-association of zwitterionic materials, and prepared and characterized zwitterionic polymers. This work explained the anti-biofouling mechanisms of zwitterionic materials and developed principles for designing new zwitterionic moieties. One moiety designed *in silico*by us was verified to resist biofouling in experiment.

My postdoctoral research in Prof. Carol Hallâ€™s group focuses on understanding and predicting nanoparticle toxicity. Excitement about the applications envisioned for nanomaterials in energy, medicine and electronics is tempered by concerns about their toxicity. As the first response of a biological fluid to the intrusion of nanoparticles is

the formation of a protein corona that determines the nanoparticleâ€™s toxicity, we take a computational approach to learn the mechanisms that govern the formation of a protein corona around nanoparticles. A coarse-grained model was developed to describe protein-protein and protein-nanoparticle interactions. Combining this coarse-grained model, atomistic simulations and QSAR modeling, we (1) explore how binding to the nanoparticle affects the structure and dynamics of a key blood protein, albumin, (2) decode the common features in the binding configurations of 2315 human proteins and gold nanoparticles, and (3) develop models that can predict protein-nanoparticle interaction energies. This work deepens our understanding of protein-nanoparticle interactions, paving the way to predict and assess nanoparticle toxicity.

Teaching Interests:

I earned my bachelor and PhD degrees in chemical engineering and have been conducting research in chemical engineering for more than 10 years. I am confident of my abilities to teach all chemical engineering courses at the undergraduate and graduate levels, and I am eager to do so. I am particularly interested in teaching thermodynamics, transport processes, process design and applied mathematics.

465464 Computationally Assisted Discovery of Well-Designed Materials for Applications to Energy, Environment, and Catalysis

Abstract:

Research Interests:

The worldâ€™s dependence on fossil fuels has led to the need for alternate sourcesÂ of energy as supplies dwindle, as well as a growing need to removeÂ harmful compounds from the air. Hydrogen energy and lithium-ion batteries are promising candidates for supplanting fossil fuels for automobile applications while novel adsorbents like metal-organic frameworks (MOFs) are promising materials for removing harmful gases. To date, my researches have been concerned with both aspects of the fossil fuel problem in keeping with my interest in achieving better performance on polymer solar cells and understanding catalysis on alkylamine functionalized silica surfaces. Regarding the lithium-ion batteries, my researches are focused on fundamentally understanding the thermodynamics and redox properties of well-designed electrode material candidates which would directly affect the battery capacity. It is emphasized that the redox properties of carbon materials such as quinone derivatives and graphenes can be strongly correlated with their structural and electronic properties. One of the highlighted observations is that the cathodic activity of the quinone derivatives during the discharging process relies strongly on the number of carbonyl groups available for further Li binding. On the other hand, my research focus in the field of the hydrogen energy is on understanding the thermodynamics of metal hydride reactions for hydrogen storage applications. Specifically, the goal is to identify thermodynamically promising metal hydride reactions from a full database of metal hydride mixtures using first-principles calculations. The large-scale screening approach ultimately provides a number of promising single-step or multi-step metal hydride reactions. My researches on MOFs are related to investigating promising MOFs for the selective capture of harmful gases as well as gas separations. Quantum mechanical methods are used to screen and assess functional groups that would be incorporated into MOF ligands to preferentially adsorb harmful gases under humid conditions. In the last research topic, my molecular dynamics simulation approach is employed to understand the aldol condensation between acetone and 4-nitrobenzaldehyde on alkylamine-functionalized silica surfaces. In conclusion, my research goal is to contribute to the community for the transition to environmentally friendly ecosystems.

Inspired by my research experience in the field of the lithium-ion battery, my future research interests are aimed at the development of methodologies for systematically designing promising electrode materials with the high redox properties as well as the high energy and power densities. A new protocol of designing hypothetical electrode materials followed by the large-scale screening approach will be employed to expedite the material-search process. The newly designed materials are expected to provide exceptional performance in four aspects: (1) structural stability, (2) desired redox property, (3) high charge capacity, and (4) high ionic diffusivity. I believe that this research will significantly contribute to the battery community for discovering completely new electrode materials.

Teaching Interests:

I describe the topics I would like to share with my students in a future position. First, my education on quantum chemistry and molecular simulation makes it possible for me to teach an introduction to theoretical calculations and simulations. I am highly interested in sharing and discussing a number of thermodynamic and kinetic problems, which should sometimes be solved using density functional theory calculations. Second, based on my experience in gas adsorption experiments on porous materials, I would like to teach topics related to an introduction to adsorption and specific techniques for measuring adsorption isotherms at the thermodynamic equilibrium state. Third, the topics arising from my research background, namely energy materials and science, alternative energies, hydrogen storage technology, metal hydride thermodynamics, lithium-ion battery technology, adsorption thermodynamics on porous materials, carbon capture and storage, solar cell technology, and metal-organic frameworks provide interesting material to teach on both undergraduate and graduate levels. In addition, the basic knowledge I gained as an undergraduate student on general chemistry, physical chemistry, thermodynamics, fluid mechanics, heat transfer, mass transfer, differential equations, numerical analysis, and transport phenomena would prove helpful for teaching basic undergraduate classes.

472231 Computationally Driven Discovery of Novel Materials for Separation and Catalysis

Abstract:

Research Interests: Theory and molecular modeling have seen tremendous growth in recent years; capitalizing on theoretical and algorithmic developments and advancements in computer hardware, applications of computational methods have begun to penetrate into all fields of the chemical sciences. In this poster presentation, I will highlight separation and catalysis applications using crystalline nanoporous materials, primarily zeolites, and show how molecular modeling can be used both to *gain a microscopic understanding into phenomena difficult to probe experimentally* and to *predict accurate data at a scale beyond the reach of the traditional trial-and-error approach*. I developed a new, transferable force field for all-silica zeolites, TraPPE-zeo, that is applicable to any type of sorbate molecule and framework structure. With the improved intermolecular potentials as well as better methodologies, it was then possible to study several complex systems involving articulated molecules and liquid mixtures, such as the multi-component adsorption of aqueous alcohol, polyol, glucose, and furfural solutions. These studies demonstrate the important effects of hydrogen-bonding, which render both the bulk and adsorbed phases highly non-ideal and are for certain molecules so strong that adsorption becomes entropically rather than enthalpically driven. In another application, the unusual adsorption and diffusion behaviors of a hierarchical zeolite were investigated, and our simulations clearly indicate regimes where the introduction of mesoporous "highways" impedes, rather than enhances, molecular transport in a material without structural defects, and provide insights into the nature of surface resistance and its roles in the practical application of hierarchical materials and ultrathin membranes. More recently, the effect of reactant chain lengths and architectures and the degrees of framework confinement on the reactivity and selectivity of zeolites as solid acid catalysts were carefully examined using first-principles calculations. Information obtained from this work provide the basis for an effort in which a large number of materials were screened with the purpose of finding better candidates for two energy-related applications, ethanol/water separation and hydrocarbon iso-dewaxing, and some of our predictions were subsequently verified by experiments. This strategy was recently

extended to account for transition-state selectivity [1], and several advancements were made to allow for highly parallel ab-initio Monte Carlo simulations, which together bring closer the possibility of using theory and computation in a predictive mode to expedite the pace of materials discovery for complicated, realistic, and industrially relevant applications.

Teaching Interests: Thermodynamics and statistical mechanics, catalysis and chemical kinetics, heat and mass transfer, numerical methods and linear algebra, computational chemistry

[1] "A method to screen nanoporous catalysts for transition-state selectivity," to be presented at 2016 AIChE Annual Meeting, Symposium on 'Catalysis with Microporous and Mesoporous Materials,' San Francisco, CA.

476096 Towards Accurate Atomistic Description of Reactive Interfaces for *in silico* design of Novel Functional Materials

Abstract:

Research Interests:

The complex interplay of chemical reactions, defect-chemistry, transport phenomena, and structural evolution at reactive interfaces underpin much of energy capture, conversion and storage by materials. A fundamental understanding of these dynamical processes at angstrom-to-mesoscopic length scales over several nanoseconds is crucial, and urgently needed to enable development of novel functional materials for various energy applications. However, development of accurate atomistic models of reactive interfaces remains a grand challenge, mainly due to the non-systematic and intuition-driven methods employed. Here, I will first present a robust, efficient, and automatic computational framework to develop reactive interatomic potentials (force-fields; FFs) that significantly advances the state-of-the-art by completely obviating the need for human intuition. Using supervised machine learning empowered by genetic algorithms, and large datasets obtained from high-throughput *ab initio* calculations, we sample the parameter landscape efficiently to obtain a highly transferable set of FF parameters that accurately capture structural, dynamical, energetic, and other physical/chemical properties. We have successfully employed this framework for a wide range of material systems, including gold, ZrN_x, CoC_x, IrO_2, and 2D tin.[1,2] Subsequently, I will demonstrate how a synergistic integration of large-scale reactive molecular dynamics (RMD), electronic structure calculations, and *ab initio* MD (AIMD) can pave path towards an accurate description of reactive interfaces, and consequently enable computational discovery of novel materials. Specifically, I will highlight two representative examples: (1) using AIMD, and million-atom RMD simulations, we discovered a novel catalytic phenomena possible only under the extreme local conditions afforded by friction at sliding interfaces; this enabled design of novel catalytically active coatings (e.g., MoN_x-5%Cu) with exceptional wear resistance (~25% lower friction than best blended oils). We found that such coatings transform base lubricating oils (long-chain alkenes) into low-friction diamond-like carbon (DLC) film via dehydrogenation, backbone scission of long chain alkenes to form short fragments, and their subsequent polymerization. Our atomistic simulations also indicate that the formation of DLC-like tribofilm is crucially linked to the propensity of the transition metal catalyst to form carbides; carbide-forming metals, e.g., V preclude the formation of DLC tribofilm.[3] (2) RMD simulations revealed a new transfer-free route to synthesize wafer-scale high quality single/multi-layer graphene directly on diamond.[4]We found that a Ni thin film placed on ultra-nanocrystalline diamond (UNCD) causes amorphization of diamond lattice via diffusion of Ni into the UNCD grain boundaries (GBs); concurrently, C from UNCD GBs diffuse through the Ni film, and form a graphene layer on the surface. This opens up avenues to explore novel synthesis pathways, and atomistic mechanisms underlying exotic phenomena in other 2D materials, e.g., stanene, phosphorene etc.

Research Experience

My research experiences lie in the broad area of theoretical/computational materials science, which have made me well-versed with developing accurate models at multiple length scales (angstrom to meso) and employing them to gain novel physical insights into exotic materials behavior. My expertise includes density functional theory calculations, *ab initio*molecular dynamics, large-scale classical molecular dynamics simulations, multi-parameter optimization for empirical force-field development, machine learning, advanced sampling and mesoscopic length-scale models. Specifically, I have employed these tools to investigate (i) structural evolution, ionic diffusion, and solvation dynamics near electrochemical interfaces during aqueous corrosion, (ii) atomic-scale diffusion mechanisms in ionic liquid based electrolytes for Al-batteries, (iii) mechanisms underlying friction and chemical reactions at sliding interfaces, (iv) growth mechanisms for 2D materials on transition metal surfaces, and (v) self-assembly of biomolecules (like collagen, peptoids) on inorganic surfaces. Furthermore, as a part of the user program at Center for Nanoscale Materials (DOE nanoscience user facility at Argonne National laboratory), I have fostered successful collaborations with several experimental research groups at US universities; Through these user-collaborations, I have worked on materials science problems at various length/timescales including investigations on large area synthesis of 2D materials via electrochemical exfoliation, chemical reactions under tribological conditions, and self-assembly of ligand decorated nanoparticles.

Postdoctoral Project

*"Bridging the electronic and atomistic scales: force field development for reactive interfaces from first principles"*Under supervision of Dr. Subramanian Sankaranarayanan, Center for Nanoscale Materials, Argonne National Laboratory

PhD dissertation

*"Understanding structure-property relationships in b-eucryptite through atomistic simulations"*Under supervision of Prof. Cristian V. Ciobanu (Mechanical Engineering) and Prof. Ivar E. Reimanis, (Metallurgical and Materials Engineering), Colorado School of Mines.

Future Direction

Despite recent strides, the field of computation-driven materials design remains plagued by the dearth of accurate models (at nano-to-meso length scales), as well as lack of efficient strategies to design new realistic functional materials (i.e, with defects, grain boundaries etc.) and identify routes to synthesize them. Motivated by this challenge, as a faculty, I seek to establish a successful research program titled "Multi-scale *in silico* materials design", which focuses on devising new strategies/tools to address these challenges, and applying these tools to investigate various technologically relevant materials systems. Specifically, I intend to develop robust computational frameworks to (1) use advanced sampling methods (e.g., evolutionary algorithms, metadynamics) to search for new metastable/stable structures for low-dimension materials (i.e., clusters, 2D sheets), (2) couple high throughput-DFT calculations with advanced global (e.g., genetic algorithms) and local (e.g., Simplex) optimization routines to develop new reactive interatomic force field (FF) models, which capture bond formation/dissociation events and transition states accurately; such FFs are crucial to advance fundamental understanding of interfacial reactions, mechanistic origins of materials phenomena, and structure-property relationships at length scales spanning tens of nanometers, (3) develop machine learning strategies to rapidly design new materials for superlubricity, battery electrolytes and anti-corrosion coatings, as well as (3) develop accurate coarse-grained models to understand self-assembly of hierarchical architectures at *mm* sizes, including superlattices of ligated nanoparticles.

Teaching Interests:

I possess a strong interest in teaching and mentoring students. During my postdoctoral appointment at Argonne National Laboratory (ANL), I pursed and obtained the opportunity to mentor several graduate students who actively collaborate with my group on various computational materials science research problems. While working towards my PhD dissertation, I delivered several guest lectures for the course "Ceramic Engineering" hosted by the Department of Materials Science and Engineering in Colorado School of Mines (CSM); additionally, I have also served as a teaching assistant for various materials science courses at CSM [e.g., Electronic Properties of Materials, Mechanical Properties of Materials, Nanomechanical Engineering etc.]. These opportunities have helped me develop the qualities necessary for effective teaching, and for being a good mentor.

Selected Publications

1. **B. Narayanan**, A. Kinaci, F. Sen, M. J. Davis, S. Gray, M. K Y. Chan, and S. Sankaranarayanan, "*Describing the diverse geometries of gold from nanoclusters to bulk – a first-principles based hybrid bond order potential*", Journal of Physical Chemistry C 120, 13787 (2016).
2. **B. Narayanan**, K. Sasikumar, A. Kinaci, F. Sen, M. J. Davis, S. Gray, M. K Y. Chan, and S. Sankaranarayanan, "*Development of a modified embedded atom force field for zirconium nitride using multi-objective evolutionary optimization", Accepted for publication*, Journal of Physical Chemistry C (2016).
3. A. Eridemir, G. Ramirez, O. Eryilmaz, **B. Narayanan**, Y. Liao, G. Kamath, and S. Sankaranarayanan, "*Cabon-based protective tribofilms from lubricating oils", In Press*, Nature (2016).
4. D. Berman, S. Deshmukh, **B. Narayanan**, S. Sankaranarayanan, Z. Yan, A. Baladin, A. Zinovev, D. Rosenmann, and A. Sumant, "*Metal induced ultra-fast transformation of diamond into single domain graphene on wafer scale*", Nature Communications 7, 12099 (2016).
5. **B. Narayanan**, S. A. Deshmukh, L. K. Shreshta, K. Ariga, V. Pol, and S. K. R. S. Sankaranarayanan, "*Cavitation and radicals drive the sonochemical synthesis of functional polymer spheres", Accepted for publication*, Applied Physics Letters (2016)
6. **(Invited) B. Narayanan**, S. Deshmukh, S. Sankaranarayanan, and S. Ramanathan, "*Strong correlations between structural order and passive state at water–copper oxide interfaces*", Electrochimica Acta 179, 386 (2015)
7. **B. Narayanan**, G. Gilmer, J. Tao, J. DeYoreo, and C. Ciobanu, "*Self-assembly of collagen on surfaces: the interplay of collagen-collagen and collagen-substrate interactions*", Langmuir 30, 1343 (2014)
8. **B. Narayanan**, A. van Duin, B. Kappes, I. Reimanis, and C. Ciobanu, "*A reactive force field for lithium–aluminum silicates with applications to eucryptite phases*", Modelling and Simulation in Materials Science and Engineering 20, 015002 (2012)

WAGENINGEN UR
For quality of life

Solubility and thermodynamic properties of α-amino acids in a model system of industrial residues

Nathan A. Bowden, Johan P.M. Sanders, Marieke E. Bruins

Background

- Amino acids have the potential to play an important part in a Bio-based Economy,
- This potential is expressed in the possibility of extracting amino acids out of agro-industrial residues for feedstock or bio-based products,
- Extracting and crystallizing single amino acids from a solution of mixtures of amino acids will facilitate reaching their economic potential in the bio-based pyramid (figure 1)

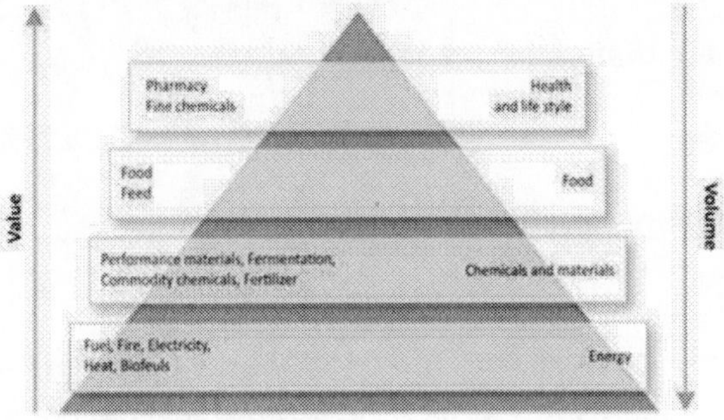

Figure 1: The bio-based pyramid

Methodology and Objectives

1. Extract individual α-amino acids from a mixture of 20 α-amino acids by crystallization,
2. Measure the solubility of α-amino acids in a model system of all 20 α-amino acids from 277.15 K to 335.15 K
3. Compare the solubility of these α-amino acids in in a model mixture to themselves in single solution

Results

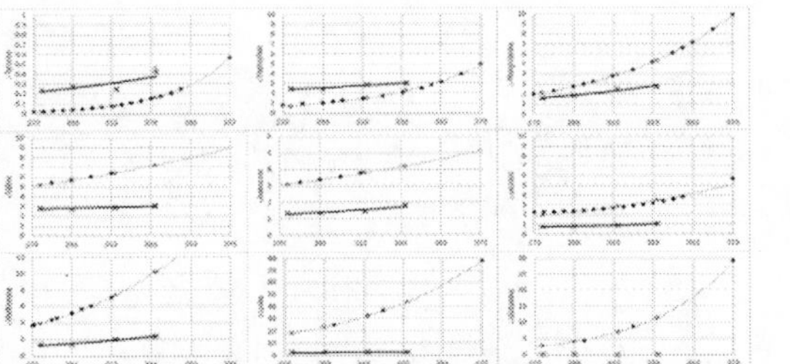

Figure 2: Solubility (m_i) of α-amino acids in g/100 g water at various temperatures (K). Solid circle (single amino acid from literature), empty circle (single amino acid calculated from literature values), solubility as a function of temperature (dotted line) crosses with error bars (measured in the model system of multiple amino acids). The solid line represents the function of the solubility of the amino acid in the model system of 20 α-amino acids.

Results (Cont.)

- L-tyrosine and L-tryptophan dissolved in higher concentrations in the model system of 20 α-amino acids than in aqueous solution by themselves,
- The solubility of the aliphatic α-amino acids (L-phenylalanine, L-valine, L-isoleucine, L-leucine and L-methionine) were lower in the model system of 20 α-amino acids,
- L-lysine and L-glutamine showed the most decrease when in the model system of 20 α-amino acids.

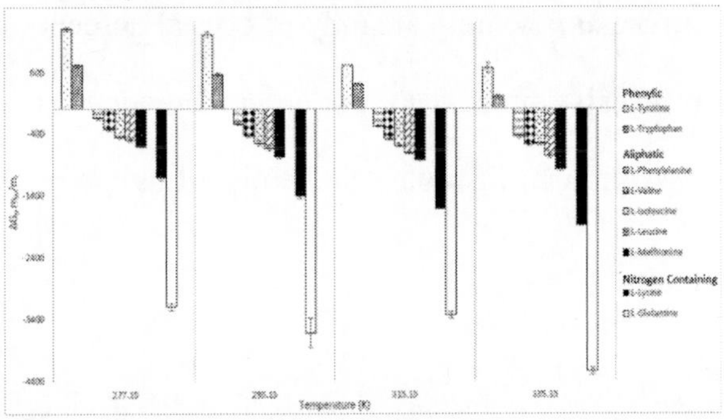

Figure 3: The standard free energy of transfers, ΔG^{o}_{tr}, for the α-amino acids grouped by their side chain physical characteristics

Conclusions

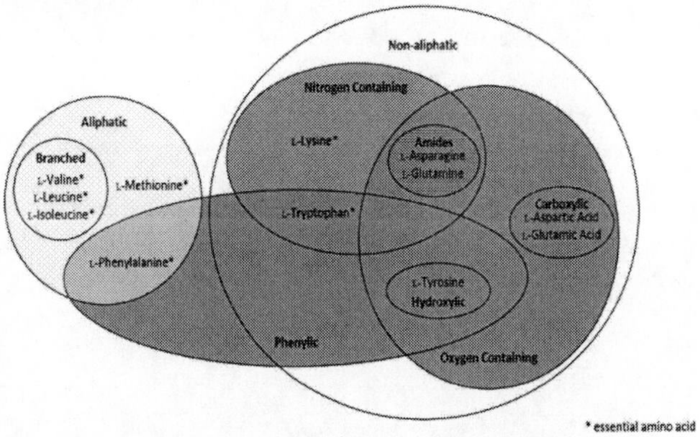

* essential amino acid

Figure 4: A Venn diagram of the physical characteristics of the side chains of 12 α-amino acids found in agricultural residue in this study. Solubility when in a mixture: increased (blue), slightly reduced (yellow) and greatly reduced (red).

- Our results demonstrate that α-amino acids behave differently in mixtures than in aqueous solution alone and since structurally similar α-amino acids behave similarly in the model system, we assert that the form of an α-amino acid predicts its solubility.

Wageningen UR, Agrotechnology and Food Sciences Group,
Sub-department of Environmental Technology
and Biobased Chemistry and Technology Group
Droevendaalsesteeg 4, 6708 PB Wageningen
Contact: elvira.bozileva@wur.nl
T + 31 (0)317 482943
www.wageningenUR.nl

avans hogeschool

453432 Dispersion-Corrected Density Functional Tighiting Binding Modeling of the Natural Metastable Twin Boundary of Organic Energetic Materials: Beta-Octahydro-1,3,5,7-Tetranitro-1,3,5,7-Tetrazocine

Abstract:

A comparative study of the geometries and electronic properties of twin boundary (TB), as a typical planar defect, in crystalline Î²-boctahydro-1,3,5,7-tetranitro-1,3,5,7-tetrazocine (Î²-HMX) was conducted by using self-consistent charge density functional tight binding with dispersion corrections (SCC-DFTB-D). The possibleÂ twinned structures were predicted and compared with many experiments. The calculated electronic properties of the proposed TBs showed that the interface of twin boundary were more active than the bulk region. The TBs grown along the (001) plane was shown to be more energetically stable than other TBs along (101) and (10-1) plane. The most stable TB structure was compared with available experimental results. The formed TB in the ideal bulkÂ increases internal instability and thus increases the sensitivities of the real materials to the external stimuli.

Research Interests: theoretical study of crystal defects of organic materials.

Teaching Interests: molecular simulation methods.

467626 Quantitatively Reliable Molecular Modeling and Simulation of Vapor-Liquid Equilibria

Abstract:

To reach the quantitative precision required in engineering applications, molecular force-field models need to be adjusted to experimental data for diverse thermodynamic properties. This implies the presence of multiple conflicting objectives in the optimization task, making it difficult to achieve an optimal agreement for the different properties with a single model parameterization. In the present work, this problem is solved by computing and analysing the Pareto set, i.e. the set of all rational compromises between objectives, from which a user can choose the one which fits best to a particular application scenario.

For a variety of low-molecular fluids, two-center Lennard-Jones plus point-quadrupole (2CLJQ) and two-center Lennard-Jones plus point-dipole (2CLJD) models are applied to describe bulk and interfacial properties of pure-component vapor-liquid equilibria. Computing the surface tension at a high accuracy by molecular dynamics simulation requires a massively-parallel simulation environment with a numerically efficient long-range correction for planar interfaces [1, 2]. For this purpose, a highly scalable molecular dynamics code is used: ls1 mardyn [3]. Literature models, which were adjusted to bulk properties, but not to interfacial properties, are validated against the surface tension of real fluids. These models, on average, overestimate the surface tension by 15 to 20% [4, 5].

On this basis, the following optimization task is studied: For ten different pure fluids, the 2CLJQ model class is explored as a whole, and by computing the Pareto set, it is evaluated to what extent the vapor pressure, the liquid density, and the surface tension can be accurately modelled by one and the same model. In various cases, it is found that any pair of two out of the three properties can be described by a 2CLJQ model with an accuracy close to the accumulated error from experiment and simulation, but it is impossible to reconcile all three properties to that level of accuracy [6].

It is computationally expensive to determine the Pareto set. Brute force methods usually cannot be applied even for moderate dimensionalities of the objective and parameter spaces, whereas Monte Carlo methods often yield poor results. For the present study, two algorithms are combined: Sandwiching and hyperboxing. This is computationally efficient and enables a specification of the numerical error, illustrating that multicriteria optimization provides a versatile framework for developing thermodynamic models. User-friendly visualization techniques for the Pareto set are discussed, including a novel approach based on self-organizing patch plots [6].

[1] S. Werth, G. Rutkai, J. Vrabec, M. Horsch, and H. Hasse, Molecular Physics 112 (2014) 2227-2234.
[2] S. Werth, M. Horsch, and H. Hasse, Molecular Physics 113 (2015) 3750-3756.
[3] A. Heinecke, W. Eckhardt, M. Horsch, and H.-J. Bungartz, Supercomputing for Molecular Dynamics Simulations, Springer, Heidelberg (2015).
[4] S. Werth, K. Stöbener, P. Klein, K.-H. Küfer, M. Horsch, and H. Hasse, Chemical Engineering Science 121 (2015) 110-117.
[5] S. Werth, M. Horsch, and H. Hasse, Journal of Chemical Physics 144 (2016) 054702.
[6] K. Stöbener, P. Klein, M. Horsch, K. Küfer, and H. Hasse, Fluid Phase Equilibria 411 (2016) 33-42.

Research Interests:

Computational Molecular Engineering (see above).

Teaching Interests:

Courses presently taught by Martin Horsch at the University of Kaiserslautern, Germany:

Molecular Thermodynamics (in German): Lecture covering statistical mechanics (ensemble, phase space, partition function, analytically tractable problems) as well as molecular modeling and simulation (Monte Carlo and molecular dynamics, computation of thermodynamic properties).

Molecular Simulation (in German): In this "computer laboratory" course, the students implement and apply molecular simulation methods under individual supervision. This is accompanied by a lecture concerning the simulation of phase equilibria (entropic quantities, Grand Equilibrium, interfacial properties) and non-equilibria (fluctuation and dissipation, transition states, work done by small systems), extending the knowledge from Molecular Thermodynamics.

Molecular Modeling (in English): Compact course within the international graduate school IRTG 2057 "Physical Modeling for Virtual Manufacturing Systems and Processes" (University of Kaiserslautern, UC Berkeley, and UC Davis), covering basic molecular modeling and simulation approaches and relating them to other methods in computational engineering.

Martin Horsch was positively evaluated as a Junior Professor in 2014, and following a training in didactics, he obtained the Rhineland-Palatinate Certificate for University Didactics in 2015. Preceding his faculty appointment, he served as a teaching assistant in Mechanical Engineering (Molecular Thermodynamics) and Computer Science (Theoretical Computer Science II and III) at the University of Stuttgart, and in Mechanical Engineering at the University of Paderborn (Thermodynamics II and Refrigeration Technology).

476178 Modeling Chemical Reactivity for Nanoscale Design

Abstract:

Research Interests: The overarching goal of my research is to understand the connection between molecular interactions and macroscopic dynamic behavior. I employ computer simulations at both classical and quantum levels of detail to probe molecular interactions, and use reaction rate theories and statistic mechanics to evaluate the related macroscopic kinetic and transport parameters. My research underscores the need for non-empirical and transferrable predictive tools for the design of new solvents and catalysts. Specific areas of interest include identification of mechanisms for solvated and heterogeneous reactions; design of solvents and solid surfaces with unique interactions that increase selectivity; development of efficient simulation methods for calculating transport properties in amorphous condensed phases with long relaxation times.

Postdoctoral Project: Reaction equilibria in ionic liquids using Monte Carlo simulations
In collaboration with Ed Maginn & Steve Corcelli, Chemical and Biomolecular Engineering, Notre Dame

PhD Dissertation: "Mechanisms of Rare Events in Condensed Phases"
Under supervision of Baron Peters & Joan-Emma Shea, Chemical Engineering, University of California Santa Barbara

Teaching Interests: I am prepared to teach any of the core chemical engineering undergraduate courses, particularly mass transfer, chemical reaction engineering, and thermodynamics. I am additionally interested in teaching introductory chemical engineering courses to first and second year students, a course on analytical mathematical methods, or a course on numerical methods / computer programming. I am eager to develop elective courses on molecular simulation at both the undergraduate and graduate levels, and graduate courses in reaction rate theory and mass transfer. I find that illustrating fundamental chemical engineering processes (e.g. diffusion, phase separation, reaction equilibrium) using visuals at both the macroscopic and molecular levels better prepares students to reason

through new scenarios. When feasible, I plan to incorporate course projects with hands-on demonstrations or molecular-level visualization in order to build an online repository of teaching aids.

476189 Enhanced Molecular Simulations for Applications in Protein Stabilization, Crystallization, and Structural Determination

Abstract:

1st Year Postdoctoral Fellow

Research Interests:

Molecular modeling and simulation can provide researchers with a more detailed understanding of the interactions that underlie important biological and chemical processes. With ever-advancing computer power, researchers can now simulate the molecular-level interactions of proteins, DNA, large polymers, and other organic macromolecules on experimentally relevant timescales. One of the major shortcomings of standard molecular simulations techniques is that they sample molecular conformations that are near local free energy minima. Thus transitions between various states of a system and the barriers associated with those states are rarely visited during a simulation. To overcome these limitations, several enhanced sampling techniques have been developed that allow a researcher to exhaustively sample states of the system and escape local free energy minima. Some of the rarely visited states are often crucial to completely understand thermophysical and transport properties, and such information can not only help us to understand chemical phenomena but also to predict the behavior of chemical systems *in silico.* I an expert in the application of enhanced and non-equilibrium sampling methods that allow us to answer pertinent scientific questions.

Successful Proposals: Alexander von Humboldt Postdoctoral Fellow. Assisted in writing and gathering preliminary results for NSF and XSEDE/Teragrid applications.

Postdoctoral Project: *Enhanced Molecular Simulation and Thermodynamic Characterization of* Protein Mutations for Improved Crystallization by Surface Entropy Reduction

Externally funded through the Alexander von Humboldt Foundation. Under the supervision of Prof. Bert de Groot at the Max Planck Institute for Biophysical Chemistry. Göttingen, Germany.

PhD Dissertation: *Developing a Simulation Framework for Modeling Biomolecules in Ionic Liquids*

Under the supervision of Prof. Jim Pfaendtner at the University of Washington. Seattle, WA.

Research Experience:

The focus of my research has been the computational modeling of protein and ionic liquid systems. As such, I have become an expert in the field of molecular dynamics simulations with an emphasis on enhanced sampling methods that allow us to characterize the thermodynamics of rare events. Specifically, I employ metadynamics and alchemical mutations to determine the relative free energies of multiple states of biomolecular systems and rates of transitions between states as in protein unfolding and ligand unbinding. When I initiate a simulation study, I always look to better explain the molecular interactions that drive experimental results or to predict results of future experiments. Other times, experimental collaborators request my assistance in performing molecular simulations to determine the feasibility of a hypothesis. As such, I have worked closely with several collaborators and even collected experimental data myself. Consequently, I have experience in performing simulations that are relevant to IR, Raman, and SFG spectroscopy as well as microcalorimetry, reaction kinetics, AFM, electron microscopy, XPS, SAXS, and SANS. Therefore, I am confident that I can design and implement simulations that (a) answer important questions about biological and chemical phenomena, (b) support the development of hypothesis in collaboration with other scientists, and (c) efficiently and quickly predict the behavior of biomolecular and organic systems when no experimental data is yet available.

Teaching Interests:

During my time as a PhD student, I had the privilege of training several undergraduate, masters, and fellow young PhD students in the fundamentals of molecular simulation. I find teaching and training of students to be one of the most rewarding parts of my academic career. As a TA, I had the opportunity to teach both core chemical engineering courses such as unit operations, mass transport, and separations as well as electives such as introductory courses to molecular and nanoscale principles. During these courses I gained experience in developing lectures and in the hands-on training of students for the operation of scientific equipment. As a professor, I would feel comfortable in teaching any core chemical engineering course, and I would like to develop a course on molecular simulation to help students better understand the fundamental physics that underlie the thermophysical phenomena we observe as scientists and engineers.

Future Direction:

As faculty, I would like to move my research forward in three distinct fields:

First, I would like to expand my work on thermostabilizing and entropy-reducing protein mutations to include alchemical thermodynamic calculations on mutations that affect the stability of proteins in the presence of chemical denaturants and of enzymes in confinement. With such simulations, we can design proteins that better tolerate a wide range of conditions. The engineered proteins are industrially relevant for a wide variety of applications where enzymes are subjected to stress as in biomass processing, food production, and long-term therapeutic protein storage.

Second, I would like to study peptides that precipitate inorganic nanoparticles. Our previous work indicates that coarse-grained metadynamics simulations can give us a general idea of the shapes of nanostructures formed by a core of small charged peptide with only 1-2 days of simulation time. In nature, many organisms use proteins to transform simple soluable molecules into complex inorganic structures such as diatom exostructures, mollusk shells, and vertebrate bones. Using natural proteins as a starting point, we can run a campaign of simulations to screen thousands of potential nanostructure-forming mutant peptides per year. This work offers the possibility to make well-defined inorganic nanostructures at mild conditions for applications such as catalyst supports or drug delivery.

Third, I would like to develop a protocol for accurately predicting the properties of mixtures of ionic liquids using polarizable force fields. Ionic liquids are liquid organic salts that have many unique properties making them interesting for industrial applications such as biomass processing, batteries, and carbon dioxide capture. Our previous work on ionic liquids has demonstrated that the thermophysical properties of many pure ionic liquids can be predicted using molecular simulation with a generalized non-polarizable force field. We can improve our results by using established polarizable force fields, and we can begin to model mixtures of multiple ionic liquids. With the right computational tools, we can develop highly-tuned ionic liquid recipes to fit the properties required for a wide range of applications.

Selected Publications:

V.W. Jaeger and J. Pfaendtner. *Structure, Dynamics, and Activity of Xylanase Solvated in Binary Mixtures of Ionic Liquid and Water. ACS Chemical Biology.* 2013. *8* (6), 1179-1186.

V. Jaeger, P.R. Burney, and J. Pfaendtner. *Comparison of Three Ionic Liquid-Tolerant Cellulases by Molecular Dynamics. Biophysical Journal.* 2015. *108 (4)*, 880 â€“ 892.

J.E. Baio, A. Zane, **V. Jaeger**, A.M. Roehrich, H. Lutz, J. Pfaendtner, G.P. Drobny, and T. Weidner. *Diatom Mimics: Directing the Formation of Biosilica Nanoparticles by Controlled Folding of Lysine-Leucine Peptides. Journal of the American Chemical Society.* 2014. *136* (43), 15134-15137.

K.G. Sprenger, **V.W. Jaeger**, and J. Pfaendtner. *The General AMBER Force Field (GAFF) Can Accurately Predict Thermodynamic and Transport Properties of Many Ionic Liquids. The Journal of Physical Chemistry B.* 2015. *119* (18), 5882-5895.

V. Jaeger, W. Wilson, and V.R. Subramanian, *Photodegradation of Methyl Orange and 2,3-Butanedione on Titanium-Dioxide Nanotube Arrays Efficiently Synthesized on Titanium Coils. Applied Catalysis B: Environmental.* 2011. 110, 6-13.

476191 Advanced Materials Design Using Molecular Simulation, Evolutionary Computing and Machine Learning

Abstract:

Research Interests: Polymer and Soft Matter; Glass Transiton; Self-Assemly, Phase Transition and Free Energy of Materials; High Throughput Materials Design; Machine Learning

Teaching Interests: Thermodynamics, Statistical Mechanics, Theory of Liquids, Polymer Physics, Numerical Methods, Molecular Modeling

Abstract: Molecular simulation has emerged as a remarkable technique for investigating structure-property relationships in a wide variety of materials. Combining it with evolution based optimization methods such as genetic algorithm can provide tremendous opportunity for discovering new materials of extremal properties. However, the major roadblock in this direction is that the determination of a molecular property via simulations is very slow. Alternatively, machine learning tools have been progressively adopted by the materials science community to accelerate design of materials with targeted properties. Nevertheless, in the search for new materials exhibiting properties and performance beyond that previously achieved, machine learning approaches are frequently limited by two major shortcomings. First, they are intrinsically interpolative. They are therefore better suited to the prediction of properties within the known range of accessible behavior than to the discovery of new materials with extremal behavior. Second, they require the availability of large datasets, which in some fields are not available and would be prohibitively expensive to produce. Here we describe a new strategy for combining genetic algorithms, machine learning, and molecular simulation to discover materials with extremal properties in the absence of pre-existing data. Predictions from progressively constructed machine learning tools are employed to bias the evolution of a genetic algorithm, with fitness evaluations performed via direct molecular dynamics simulation. We survey several materials design problems, in particular, polymer compatabilizer, porous polymeric materials, and polymer glasses, with in this framework. Further development of this computational framework and its implication to future material design will be discussed in this presentaion. In additon, we employ molecular simulation to provide useful guidelines for designing polyelectrolytes and polymer nanocomposites.

476138 Exploring Fundamentals of Zeolite Catalysis â€“ a Theoretical Perspective

Abstract:

Research Interests:

Zeolites are among the most important industrial catalysts today. Their diverse structure leads to a large variety of different possible active sites combined with a high selectivity due to their microporous structure. However, chemical reactions take place inside the material, which makes a full understanding of them highly challenging. This is one of the reasons, why many fundamental aspects of zeolite catalysis are still little understood. In this contribution I describe my efforts in the past years to elucidate several of the underlying principles in context of several timely problems within zeolite catalysis.

One of the main challenges in automotive catalysis using lean burn engines is the removal of nitrous oxides from exhaust gases. A system that has been found to be a particularly efficient catalyst in this context is copper exchanged SSZ-13 (Cu-SSZ-13). Even though SSZ-13 is a zeolite in the chabazite structure, the zeolite structure with the most simple primitive unit cell, the true nature of the active centers for this reaction is still under discussion. In this contribution I describe my efforts to combine ab-initio molecular dynamics simulations, post Hartree-Fock calculations and thermodynamic modeling to arrive at an unambiguous assignment of infrared spectra of adsorbed probe molecules [1,2] and UV-vis spectra of reduced Cu-centers [3]. Due to the thermal motion at finite temperature, the Cu cation is highly mobile within the framework. This mobility changes with the oxidation state. While Cu^{II} oscillates around its local minimum, Cu^{I} changes its coordination on a pico-second timescale [1,2,4] and Cu^{0} even moves from its favored position in the six-ring of the structure to silanol-defect sites [3]. However, as soon as water is present in the system, the cation is solvated and moves away from the framework. While it is possible to follow this mobility by comparing calculated spectra to experimentally measured ones, it is still exciting to see in what way these fundamental insights on the nature of the active sites will affect our understanding of the reaction cycle and whether they can be transferred to other zeolite catalyzed reactions.

Another reaction that has recently drawn large amounts of attention is the conversion of methane-to-methanol over transition metal exchanged zeolites. First of all these systems show a very low activation energy and at the same time the active centers are very similar to those in enzymatic catalysts, responsible for catalyzing this reaction in nature. Here I will discuss two different aspects of this problem. It is safe to assume that the confining environment is a key factor for the activity of these materials. Confinement effects are rather complicated to understand, since entropy and enthalpy affect adsorption and conversion of reactants and products. A first step is understanding the impact of confinement on the activation enthalpy of the methane-to-methanol conversion over Fe-oxo centers in SSZ-13 [5]. I conclusively show that confinement stabilizes the reaction intermediates more than the initially adsorbed methane molecule, which in turn leads to a significant lowering of the activation energies. This allows to extrapolate generally valid statements about the impact of confinement on reaction enthalpies, which will most likely impact our understanding of similar processes in the future. However, more recently Cu-exchanged zeolites have drawn larger amounts of attention in the scientific community for this reaction. I will present a thermodynamic model for $Cu_xO_yH_z$ sites in the zeolite SSZ-13 [6]. Interestingly not only one, but several different types of active sites exist in this material and depending on the exact conditions, different ones are thermodynamically most stable. The theoretical model is also supported by Raman- and UV-vis measurements, which show excellent agreement with the modeled results.

The problems discussed in this contribution show, how of state of the art electronic structure theory combined with experimental measurements can help elucidating fundamental concepts in zeolite catalysis. In the future I am looking forward to continue my work along this path to improve the conceptional understanding of zeolite catalysis, to generalize these ideas and apply them to catalysts of industrial interest and ideally arrive at an approach that does not only help to understand, but also predict the activity of a large number of zeolite catalysts. Furthermore I belive that collaboration is one of the key aspects of succesful scientific work and I am looking forward to establish joint projects with other faculty members.

Teaching Interests:

Besides my achievements in science, I have always had a passion for learning and teaching, which has been reflected in mentoring graduate and undergraduate students as well as teaching selected classes during my current post-doc. In my time as a faculty member I plan to develop a course describing how state of the art electronic structure methods can be applied to gain insights into heterogeneous catalysis, either by understanding reaction pathways or by predicting observed spectra. It will provide the necessary theoretical background as well as practical examples and hands-on experience for the students. Among the undergraduate core classes I ready to teach thermodynamics and, depending on the departmental needs, am looking forward to develop classes close to my core competencies in fundamental physics, quantum mechanics, catalysis or materials science.

476139

Abstract:

Reverse Engineering of Molecular Structure

Dr. Farhad Gharagheizi

Email: fghara@gmail.com ; farhad.gharagheizi@ttu.edu ; Tel: (+806) 252 8892

Post-Doctoral Researcher (Advisor: Ronald Hedden)

Department of Chemical Engineering, Texas Tech University,

Box 43121, Lubbock, Texas 79409-3121, United States

Advanced computational techniques are invaluable for investigating how the molecular structure of pure organic/inorganic compounds, ionic liquids, polymers, and their mixtures affect their physical, chemical and biological properties. Molecular structure-based knowledge gained from computational screening can be harnessed to reverse-engineer new chemical compounds with desired properties. Molecular structures of chemical compounds possess encoded details that govern their thermophysical properties. Decoding these structural characteristics through data-driven computational screening enables reverse engineering of molecular structure to obtain molecules with desired properties. Reverse engineering methodology is relevant to research in the fields of Petroleum Engineering, Natural Gas Science and Engineering, Pharmaceutical Science, Environmental Science, and Energy.

In the petroleum field, reverse engineering of molecular structure could be used to design new materials that have attractive transport properties (such as viscosity, thermal conductivity, heat capacity, density, and vapor pressure). As an example, one could target the design of a "perfect" lubricant whose viscosity has minimal temperature dependency within a desired range of temperature and pressure. In natural gas science and engineering, reverse engineering of molecular structure could also be used to design safer separation media. For instance, the technique could be used to find a new chemical compound that promotes the storage capacity of gas hydrates and improves process performance.

In the energy area, electrolytes play a key role in energy storage systems (i.e. batteries, super-capacitors, actuators, dye sensitized solar cells, and thermo-electrochemical cells). Also, the tunable phase behavior of ionic liquids also permits the creation of phase-change thermal energy storage materials having melting points tuned to any desired application. Reverse engineering of molecular structure could be used to design new ionic liquids that exhibit attractive combinations of viscosity and conductivity.

In the pharmaceutical industry, “Drug Discovery” generally refers to a high-throughput screening process in which the main goal is to find a chemical compound that meets predefined performance criteria. Such criteria are generally defined based on the values of a series of properties/activities (i.e. aqueous solubility (blood solubility), skin permeation, organ diffusivity, etc.). Therefore, molecular reverse engineering could play a key role in identifying undiscovered drug molecules with structural features that induce the desired performance characteristics.

As an assistant professor, I plan to establish a mainly computational research group in molecular engineering, develop strong collaborations with industry, and serve as a key element of interdisciplinary research teams with experimentalists and theoreticians.

- Abstract Text

476146 Developing Molecular Theories/Simulations to Understand and Optimize Soft Matter Systems: From Ions to Polymers to Gels

Abstract:

Research Interests:

Modern chemical engineering becomes increasingly molecular-based. The development of theoretical frameworks that capture the essential physics, address the molecular feature and quantify the systems across multiple time and length scales is vitally important. My research focuses on using statistical mechanics, which bridges the molecular information and macroscopic phase/interfacial properties, to study a variety of soft matter systems including electrolyte solutions, liquid crystals, polymers, polyelectrolytes, polymer-protein conjugates and polymer networks/gels.

I did my PhD work at Caltech with Prof. Zhen-Gang Wang, where we developed renormalized Gaussian fluctuation theory to study the effect of self energy of ions on the double layer structure and interfacial properties. Our theory incorporates different components of self energy in a unified framework and involves only intrinsic parameters. We show the consequence of the self energy by revealing three essential effects: the image charge effect, inhomogeneous screening effect and specific ion effect. We are able to quantitatively explain a couple of long-standing puzzles in physical chemistry and soft matter physics, such as the salt concentration dependence of surface tension, like-charge attraction and charge inversion, Hofmeister series in the interfacial affinity and strong interfacial adsorption of hydrophobic ions.

My current postdoctoral research in MIT with Professors Bradley Olsen, Alfredo Alexander-Katz and Jeremiah Johnson is to use theory and simulation to quantify the topology and elasticity of polymer networks by addressing cyclic defects (loops). I develop a kinetic graph theory which demonstrates a universal dependence of loop fractions on the preparation condition and reveals the intrinsic relation between different cyclic topologies. I also develop a real network theory which quantifies the individual negative impacts of different loops on the elasticity. These theories shows excellent agreement with experimental data, providing, for the first time, a quantitative understanding of gel elasticity based on molecular details. By bridging the topology and elasticity, we are now seeking to optimize the mechanical property of the polymer networks through controlling the molecular connectivity.

In the future, I would like to carry theoretical and simulational research to solve fundamental challenges in physical chemistry and soft matter physics. I will build simple models that capture the essential molecular characteristics (constitution, interaction and connectivity), which are then incorporated into the framework of statistical mechanics and further solved by using or developing modern analytical, numerical and computational methods This methodology facilitates the study of inhomogeneous systems, fluctuation effects and nonequilibrium processes, where most of the challenges exist. On the other hand, I will seek for extensive collaboration with experimentalists to use this fundamental knowledge for rational design of new materials and processes. I will initially focus on the following topics:

1) Complex interfacial and kinetic phenomena driven by electrostatics;

2) Conformation, association and interfacial behavior of polyelectrolyte solutions;

3) Topology of polymer/polyelectrolyte networks vs. phase/interfacial behaviors.

Teaching Interests:

Teaching is extremely important for the professional development of my future students, and also benefits my own research work. In my future course, I would like my students to obtain clear basic concepts, the ability of critical thinking, the training of solving real problems and the knowledge about current challenges in this field. To achieve this goal, I will organize the course materials and develop the teaching methods by addressing the following aspects: 1) encourage the students to think about one problem from different perspectives; 2) guide the students to find the underlying physics that explains different phenomena; 3) design problem sets and term projects from recent research progresses.

As a graduate student at Caltech, I had the opportunity to work as a teaching assistant in graduate level thermodynamics, statistical mechanics and polymer physics. The evaluation from the students shows that I was one of the best teaching assistants in recent years. Besides these courses, my education and research experiences also make me well quantified to teach kinetics, transport process, interfacial phenomena, mathematics in chemical engineering and computer modeling and simulation. Currently as a postdoc in MIT, I am also training for mentoring undergraduate and graduate students, which is extremely helpful for my teaching skills. The course about soft materials right now is limited due to the rapid progress in this field and the lack of systematic textbooks. I plan to develop an in-depth course that convers a broad spectrum of soft matter systems in the future.

476165 Simulation of Selectively Permeable Novel Polymeric Membranes

Abstract:

My research focuses on the use of computational modeling to understand systems important in energy and materials sciences. Access to cheap, reliable energy is a grand development challenge, while simultaneously the transition to clean energy sources is a grand sustainability challenge. Often, understanding processes in energy generation requires knowledge of the physical chemistry of reactions, which occur at the atomic level on the picosecond time scale, yet also being able to track energy inputs and outputs for the total macroscopic device over the course of days or even years. Having experience with a variety of modeling methodologies that address different time and length scales (density functional theory, molecular dynamics, and stochastic kinetic algorithms) uniquely positions me to address these types of systems.

Presented here, my current project at the Joint Center for Artificial Photosynthesis (JCAP, a DOE Energy Innovation Hub supported under Award Number DE-SC0004993) is aimed at improving design and understanding of novel polymeric separations membranes. The overall goal of JCAP is to develop and demonstrate a device that can convert carbon dioxide and water into useful hydrocarbon fuels using solar energy. A key determinant of performance for an artificial photosynthesis device will be the ability to block transport of products of the carbon dioxide reduction reaction (e.g., methanol, formate) while allowing ion transport (via the bicarbonate ion). Current commercially available membranes are not designed for this purpose, and the key structural features that would enable high function for carbon dioxide reduction systems have not been identified. I am performing mesoscale simulations to develop a better understanding of key membrane characteristics. Using stochastic kinetic algorithms, I model how solutes in a complex solution move into and through polymeric materials. I work closely with experimentalists to explain experimental observations and elucidate guiding principles for design of novel membrane materials. Improvements in membrane design could have impacts in a wide range of technologies including reverse osmosis and fuel cells.

Research Interests:

My overarching vision is to improve computational modeling in order to address problems in development and sustainability. Specifically, I would (1) build on my graduate work in development of advanced force fields to expand the types of systems which can be modeled accurately with molecular dynamics, (2) extend my modeling of permeation of polymeric material to better design the controlled release of drug molecules in the body, and (3) take advantage of the long time scales that can be simulated using stochastic algorithms to understand aging of materials. Connection with experiment is and will continue to be a vital part of my research.

Teaching Interests:

As a secondary school teacher in Tanzania and then a mentor to undergraduates during my doctoral and post-doctoral work, I have found that being able to relate in-classroom knowledge to the world at large is the best way to engage students with material and have them retain the information. As such, my preferred teaching style is project-based learning, especially for upper-level classes in physical chemistry. While most class time should be spent on ensuring that students have a firm grounding in the fundamentals, they would also work on a semester-long overarching project. Also related to my interest in project-based learning, a key component of my teaching would be to involve undergraduate students in research.

476299 Towards More Rational Design of Electrocatalyst for Carbon Dioxide Reduction

Abstract:

Research Interests:

The energy appetite from our global society to satisfy the requirements of mankind's development is insatiable. However, most of the present fossil fuels (coal, oil, and natural gas) are not renewable. For the sustainable development of our society, we thus have to increasingly tap into alternative energy resources, such as wave, geothermal, solar, and wind sources, to satisfy our energy needs in the near future. In addition, the emission of carbon dioxide is unavoidable in the case of fossil fuel combustion, which results in global warming and other environmental issues. Therefore, discovering alternative energy sources and solving the environmental issues of carbon dioxide emission become more and more important. This work aims to design highly efficient catalysts for energy storage and conversion based on an innovative paradigm, i.e. theoretical descriptors based catalyst design, to solve the energy problems mentioned above. We find the binding energy of CO on electrode is an important descriptor for the activity of CO_2electroreduction.

Teaching Interests:

1. theoretical descriptor of catalysis,

2. energetic scaling relationship,

3. potential-dependent electrochemistry

476192 Computational Soft Matter

Abstract:

Research Interests:

My research aims to answer the questions: What are the molecular foundations of soft materials and how can simulations guide the design of such materials? How can we predict the assembly of novel forms of colloidal matter, based on entropy and shape? How can we design novel, biological materials from protein building blocks using simulations, and unravel the role of anisotropic shape and specificity of interaction? How can we simulate and contribute to the understanding of molecular recognition processes of proteins that are essential to the design of drugs?

I will address these questions in a vibrant research group by combining high-performance simulation on Graphics Processing Units (GPUs), theory, and coarse-grain modeling. We will perform Molecular Dynamics (MD) and Monte Carlo (MC) simulations of colloidal and biological matter, predict the assembly of proteins into aggregates and crystals, and collaborate with experimental groups on the design of protein-based materials.

Teaching Interests:

I want to convey to students the essential skills they need to participate in scientific discussion by encouraging active problem solving, presentation of research papers in small learning circles, and individual small research projects, as well as foster an inclusive learning environment. Moreover, by motivating them to arrive at solutions to problems that have multiple possible correct answers, my goal is to prepare the students for problems in research and the professional world they have not been directly prepared for. As it relates to soft matter theory, students should be able to appreciate the intellectual adventure that is science and engineering, by having a solid foundation both in classical thermodynamics as well as modern computer simulations. Ultimately, I aim to incite curiosity in the same questions that inspired the science that we teach, which I view as the best foundation for research and professional success.

476210 Design and Discovery of Multifunctional Nanoporous Materials

Abstract:

Rapid developments in design and synthesis of nanoporous materials have revolutionized both established and emerging fields including heterogeneous catalysis, gas separations, energy transformations and chemical sensing. In particular, zeolites, metal-organic frameworks (MOFs) and porous organic polymers (POPs) have received significant attention.

Given the diversity of available structures and their possible variations, atomistic simulations have proven to be an invaluable tool for understanding and predicting various properties in nanoporous materials. In this work, I will summarize several classical and quantum chemical modeling strategies for predicting reactivity, dynamics and adsorption characteristics of zeolites and MOFs. Further, I will show that these multi-scale approaches enable efficient screening of catalysts and adsorbents for industrially-relevant methane upgradation, CO_2 adsorption and olefin/paraffin separation applications.

Ultimately, as the viability of a functional material depends on a number of factors such as ease of synthesis and functionalization, long-term stability, degree of disorder and reactivity; computational techniques that transcend different time and length-scales are necessary. This work highlights the growing importance of molecular simulations, which will ultimately lead to rational design and discovery of complex, multifunctional porous materials vital for a sustainable future.

Research Interests: Nanoporous materials, multiscale molecular simulations, computational catalysis

Teaching Interests: Chemical engineering fundamentals, catalysis and reaction engineering, computational chemistry, adsorption and separations

476228 Hydrodynamic Model of Complex Liquids with Microstructure

Abstract:

Complex liquids, such as liquid crystals (LCs) and colloidal suspensions, have unique properties that find wide applications in mechanical, chemical and material engineering areas. How the underlying microstructure of such material determines its mechanical, optical and electromagnetic characteristics is not fully elucidated. The inverse problem, that what specific microstructure is needed to satisfy certain material properties, is however fundamentally important for the design of next-generation functional, active and smart materials. The goal of my proposed research is to develop and generalize continuum and molecular models to understand the physics of known experiments on structured liquids and propose the recipe of materials that can be useful for the future energy and bio-inspired material applications.

Specifically, I plan to model and simulate the following systems:

1) LC smectic phase. Highly periodic self-assembling materials are important for optoelectronics and nanotechnology applications. Focal conic domains (FCD), a typical periodic structure formed from smectic A phase when strongly confined, is the candidate for such materials. I plan to use a Q-tensor model of Landau-de Gennes free energy funtional to simulate the 3D structure of FCD. We can control the spatial configurations of FCD by introducing patterned surface. As a template, FCDs may selectively trap nanoparticles, which then show highly ordered self-assembling structure. My work should help design new periodic patterns for the future 3D electronic devices.

2) Biopolymer networks. Active nematic liquids, e. g. microtubule-kinesin-ATP film exhibits many peculiar non-equilibrium behaviors, such as active turbulence at very low Reynolds number, oscillating defect structure when

confined within a shell, and long-range orientational order among +1/2 and -1/2 defects. The underlying physics is poorly understood. I plan to develop both continuum and molecular models to explain the above phenomena at different length scales. Validated by reproducing known phenomena, my model can help understand such active materials and may guide the design of new bio-inspired materials.

3) Dense suspensions. They are important for petroleum industry, food manufacture and functional material design. Current research is focused on steady state rheology, spherical particle shape and single length scale. I plan to generalize the current simulation method, Stokesian dynamics, to account for the transient dynamics as well as multi length scale particles. My model should help understand the underlying physics of dynamic jamming, and enable the simulation of impact mitigation applications.

During my Ph. D study, I worked with Prof. Joel Koplik at Levich Institute in CCNY, on multi scale simulations of microfluidic phenomena. During my postdoc experience with Prof. Juan de Pablo at the University of Chicago, I developed the hybrid lattice Boltzmann method to simulate flowing LC system in the continuum level, and applied to the microtubules system to understand the active nematics phenomena.

Research Interests:

Theory and modelling of soft matter physics:

- Liquid crystals and its hydrodynamics

- Active nematics

- Colloidal self-assembly

- Dense suspensions

Simulation expertise:

- Molecular Dynamics

- Lattice Boltzmann method

- Stokesian dynamics

Teaching Interests:

- (Computational) Fluid Mechanics

- Statistical physics

- Molecular modelling

- Transport phenomena

- General physics

476321 New Generation of Polarizable Reactive Force Fields for Multiscale Simulations of Complex Materials

Abstract:

Research Interests:

Critical to the development and manufacturing of new generations of functional materials is the design of the non-equilibrium dynamical processing to achieve desired microstructures and transport properties. Here, we need methods for *in silico* designing and simulating of such processes to produce novel functional materials and to facilitate technological breakthroughs. These methods must be able to describe: I) non-equilibrium transport phenomena (e.g. of charges, ions, atoms, mass, and defects) in bulk and across interfaces over multiple length and time scales, II) the dynamics of phase changes during material growth, processing and operation, and III) the microstructures of complex heterogeneous structures that provide the desired functional properties.

Non-equilibrium phenomena determine the behavior of materials for practical applications, and understanding them is critical to optimize material synthesizability, characterization, and design trade-offs. Specific examples of non-equilibrium in material systems include ionic and electronic conductance in semiconductors, synthesis of thin films and nanostructures (e.g. crystallization), and functional self-organizing soft materials (e.g. adsorption polymers, catalyst support materials, soft magnets), condensed phase reaction kinetics. In all these materials, non-equilibrium dynamics determines all structural characteristics and their properties.

We need methods that provide the accuracy of Quantum Mechanics (QM) for reactive non-equilibrium dynamics simulations of extremely large spatial scales (1 billion atoms to describe a system of 100 nm on a side) and time scales (milliseconds to seconds) to characterize the non-equilibrium dynamical processes required to synthesize complex chemical and material systems. At this level we need to match to the continuum simulations required for developing manufacturing processes.

Successful Proposals: Department of Energy, National Science Foundation, Office of Naval Research

Postdoctoral Projects:

1. "New Generation Reactive Force Fields based on Valence Bond Concepts with Polarized Charge Distributions (VaPoX), Fundamental Concepts and Applications"

2. "Quantum Optimized Polarizable Charge Equilibration (PQEq) For Predicting Accurate Charges in Molecules and Solids

3. "Prediction of Structures and Properties of Green Energetic Materials from Density Functional Theory (DFT) and Reactive Molecular Dynamics"

Advisor: Prof. William A. Goddard III, Division of Chemistry and Chemical Engineering, California Institute of Technology (Caltech)

PhD Dissertation:

"A process-based molecular model of silicon carbide (SiC) nano-porous membranes"

Advisors: Profs. Muhammad Sahimi and Theodore T. Tsotsis, Mork Family Department of Chemical Engineering and Materials Science, University of Southern California (USC)

Research Experience:

My research focuses on first principles-based multiparadigm, multiscale strategies for simulating complex materials processes. It contains, 1) development of advanced computational algorithms for atomistic modeling, and 2) their applications to current problems in chemistry and chemical engineering. For example, I have developed VaPoX, a new generation of polarizable reactive force field. VaPoX allows an accurate description of the local polarization of atoms needed to describe dynamic dielectric properties. In addition, it uses valance bond concept to accurately describe the bonding. VaPoX is a great deal of progress toward *in silico* design and simulation of non-equilibrium dynamical processes required to manufacture complex functional materials to achieve novel functional and transport properties.

These research projects bring together concepts in physics, chemistry, chemical engineering, numerical methods and algorithms analysis that I have developed during my doctoral and post-doctoral work studying chemical phenomena. I have received several awards for excellence in my research including the University PhD Achievement Award (USC 2013), the Material Engineering and Science (MESD) Award (AIChE 2012), the Soft Matter Journal Award (AIChE 2012), the Mork Family Department of Chemical Engineering & Materials Science Best PhD Thesis Award (USC 2013), and Project Navigator Ltd. Environmental Engineering Awards (USC 2011 and 2012).

Future Plans:

There is an economic and environmental imperative to develop new and alternative energy technologies. As a professor, I would like to continue developing polarizable reactive force fields to design new functional materials and to facilitate technological breakthroughs.

I propose to build a new formulation of polarizable reactive force fields based on my previous successes (e.g. VaPoX) but with the goal of matching the highest level QM while making it more systematic so that the transferable reactive force fields will be attainable. This will enable not only my group but also a broad range of scientists to study nonequilibrium, adiabatic and non-adiabatic, reactive dynamics simulations of large-scale functional material systems. I would like to prove that it is possible to construct a generic force field capable of describing complex reaction dynamics at nearly the accuracy of QM on systems that are many orders of magnitude larger than QM size and time-scales.

My research direction is motivated by a diverse set of applications including energy technologies, nanotechnology, biomaterials, catalysts, and organic materials where experimental approaches are inaccessible and/or inefficient. In particular, I would like to develop new polarizable reactive force fields to design and study high energy density materials under extreme conditions, nonporous membranes for gas separation, energy storage and conversion, hydrogen storage and supercapacitors, gas purification, fuel cells, solar cells, and Li–ion batteries.

As a faculty, I hope to educate new engineers and establish an externally funded and internationally renowned research group that develops energy technologies. My research group will focus on socially important problems and the translation of scientific studies into practical technologies. I will foster a learning environment and will actively maintain collaborations with colleagues from within the engineering community and from diverse fields. Scientific advancement is best accomplished through efficient communication of results and analyses; as such, I will strive to disseminate my works at the highest scientific levels as well as to the broader public. I am confident in my abilities to manage a research group and I am highly motivated to begin working on the development of new functional materials and energy technology.

Teaching Interests:

I have experience teaching and mentoring students in the classroom and in the laboratory. As a teaching assistant I have given lectures and tutorial sessions in Transport Phenomena (i.e. Mass Transfer, Heat Transfer, and Viscous Flow), Reactor Design and Kinetics, and Advanced Engineering Mathematics. My teaching efforts and strategies received commendable reviews and ratings at USC. I received university outstanding teaching assistant award in the category of chemical engineering (2012) and Mork Family Department of Chemical Engineering and Materials Science best teaching assistant award (2013). In addition, I have received several offers from prestigious institutions such as Harvey Mudd College to teach undergraduate chemical engineering courses. I am enthusiastic about the opportunity to teach undergraduate and graduate courses including Quantum Mechanics, Computer Simulation and Modeling, Transport Phenomena, Chemical Reaction Engineering, Engineering Thermodynamics, and Advanced Engineering Mathematics.

477817 Flexible and Dynamic Porous Crystals

Abstract:

Research Interests:

1. Fundamentals of the thermodynamics of porous solids with dynamic constituents

Some porous materials possess dynamic constituents that respond to guest molecules and external stimuli. Such enzyme-like complexity can enhance the selective recognition and capture of molecules. I propose to construct and analyze simple, yet illuminating statistical mechanical models to study porous crystals with dynamic constituents to harness them for gas separations, chemical sensing, drug delivery, and catalysis.

2. Computationally-accelerated discovery of flexible materials for chemical sensing

Using a Materials Genome approach, I propose to explore the chemical space of gate-opening materials by constructing structure models from judiciously chosen molecular building blocks and carrying out high-throughput computational screenings to identify candidates for sensitive and selective chemical sensors.

3. Mechanisms of bistability and pattern formation in dynamical systems

I propose to develop mathematical models to elucidate the mechanism of the shifts in the microbial composition in the mucus layer of coral that are involved in bleaching, the water-biomass feedback involved in the self-organization of vegetation in arid landscapes, and the organization of proteins that mediate the formation of invadopodia in invasive cancer cells. These projects stem from my background in applied mathematics, but involve striking parallels with phase transitions in chemistry and reaction modeling and fluid dynamics in chemical engineering.

Teaching Interests:

Statistical Mechanics, Transport Phenomena, Reaction Engineering, Thermodynamics, generally any course on mathematical or numerical methods, data analysis, data visualization, or computer programming

476231 Computational and Experimental Investigation of Membrane Biomechanics

Abstract:

Research Interests:

Potentially toxic drugs and compounds from endogenous or exogenous sources constantly interact with the human body. Many of these compounds are small molecules that alter the physicochemical properties of physiological membranes, including changing permeability, fluidity, packing, electrostatic potential, rheology, and barrier function. In many instances, this may lead to the development of various pathophysiological states. The molecular mechanisms underlying these changes and their effects on proteins and other membrane structures are not entirely clear and depend strongly on the chemical identities of the offending compounds. My research interests focus on elucidating these molecular mechanisms through a combination of computational and experimental techniques and revealing their role in pathogenesis.

As a faculty member, I will build upon the experience and skills I have gained during my postdoctoral and graduate school training. In my lab, I look forward to applying a combination of: (i) coarse-grained molecular dynamics simulations, (ii) atomistic simulations, (iii) spectroscopic methods, and (iv) cell biology techniques. By utilizing these methods I aim to elucidate the mechanisms underlying the biophysical and biomechanical consequences of molecules involved in various pathophysiological states, leading to cardiovascular, metabolic, and immunological diseases. My research expertise, experience authoring scientific publications and writing grant proposals, as well as my experience mentoring both graduate and undergraduate students will enable me to establish a competitive research program that focuses on topics at the interface of Chemical Engineering and Biomedicine.

In my current position as a National Institutes of Health (NIH) Pulmonary T32 Postdoctoral Research Fellow at the University of Illinois College of Medicine, under the supervision of Dr. Irena Levitan, I am investigating the influence of bioactive oxidized species on model bilayer and cell membrane biomechanics. These oxidized species have been implicated in the development of atherosclerosis. In this role, by utilizing a combination of experimental and computational methods, I aim to uncover the molecular mechanisms underlying the paradoxical impact of pro-atherogenic compounds on endothelial cell stiffening and membrane biomechanics.

As a PhD student in the Chemical Engineering Department at the University of Illinois, Chicago, I undertook coarse-grained molecular dynamics simulations to probe the effects of small molecules on the structural and dynamic properties of phospholipid layers. Specifically, I investigated the membrane altering capability of toxins as applied to: (i) the differential cytotoxicity of anticoagulant species, (ii) the role played by endogenous metabolites in causing consciousness impairment, and (iii) the association of various drugs with lipid emulsions acting as detoxification agents. Using my simulation approach, I was able to obtain detailed molecular-level information on the absorption, transport, and partitioning of toxin molecules into model membranes, which provided useful insight and predictions.

Teaching Interests:

In addition to my research experience, I have extensive teaching experience, from elementary through college level. My teaching interests are Numerical and Computational Methods in Chemical Engineering, Biochemical Engineering, Material & Energy Balances, Transport Phenomena, Thermodynamics, and any other special topics related to my research background. Overall, my expertise in interdisciplinary science will serve as a framework for innovative study using both computational and experimental techniques.

476244 Machine Learning and Molecular Dynamics Map Conformational Landscape of µ-Opioid Receptor

Abstract:

Research Interests:

The µ Opioid Receptor (µOR) is a key mediator of pain and critical target for analgesic medications. The current generation of opiates – drugs that bind to µOR – are beset by dangerous side effects such as respiratory depression and addiction. In this work, molecular dynamics (MD) simulations are deployed to generate a high-resolution map of the conformational landscape of µOR. By seeding such simulations from the two X-Ray crystal structures of µOR, we not only identify key intermediates between these two structures, but also reveal new metastable states that are unforeseeable from crystallography alone. The application of Markov State Models with machine learning demonstrates tight coupling between the orientations of opiates in the binding pocket with downstream changes in the receptor's overall conformation. This clearer model of µOR dynamics may be useful to medicinal chemists in developing opiates with a more desirable therapeutic profile.

Teaching Interests:

481318 Monte Carlo Molecular Simulations of Water/Oil Interfacial Tension at Elevated Temperatures and Pressures

Abstract:

Monte Carlo molecular simulations in the NpNAT ensemble are performed to calculate the interfacial tension of water/n-dodecane, water/toluene, and water/(50wt% n-dodecane+50wt% toluene) at 1.83 MPa and temperatures ranging from 383.15 K to 443.15 K. Interfacial tension simulations of water/n-decane and water/benzene are carried out to benchmark force fields. The combination of the TraPPE forceeld and TIP4P/2005 water model overestimates the interfacial tension of water/n-decane and water/benzene at high temperatures. Modications of Lennard-Jones cross-interaction parameters are made to yield good agreement between experimental and simulated interfacial tension of water/n-decane and water/benzene.

481319 Interfacial Tension Prediction of Water/Oil Mixtures Using Stabilized Density Gradient Theory and Using iSAFT Density Functional Theory

Abstract:

The recovery eciency of crude oil from the reservoir is greatly aected by the interfacial tension(IFT) between water and oil. Large value of interfacial tension would result in oil being trapped in the porous media and hinder recovery eciency. As a result, various enhanced oil recovery techniques including surfactant ooding, steam ooding, etc, have been developed, aiming to reduce the interfacial tension between water/oil and increase the oil production. To better understand the dependence of IFT on temperature and pressure and predict IFT in reservoir conditions, stabilized density gradient theory and interfacial Statistical Association Fluid Theory(iSAFT), one variety of classical density functional theory(DFT), have been applied to study the IFT between water/dodecane, water/toluene, and water/toluene/dodecane at elevated temperatures and pressure. This work contributes to the 9th industrial challenge of predicting water/oil IFT.

481320 Prediction of the Water/Oil Interfacial Tension from Molecular Simulations Using the Coarse Grained SAFT- γ Mie Force Field

Abstract:

This work is framed within the Ninth Industrial Fluid Properties Simulation Challenge, with the aim of assessing the capability of molecular simulation methods and force fields to accurately predict the interfacial tension of oil-water mixtures at high temperatures and pressures. The current challenge focuses on predicting the liquid-liquid interfacial tension of binary mixtures of dodecane + water, toluene + water and a 50:50 (wt%) mixture of dodecane + toluene in water for a range of temperatures and pressures. In our methodology, we use coarse-grained intermolecular models with a top-down technique where an accurate equation of state is used to link the experimentally observed macroscopic properties of a fluid to the force-field parameters. The state-of-the-art version of the statistical associating fluid theory (SAFT) for potentials of variable range as reformulated in the Mie incarnation is employed here. Interfacial tensions are calculated through a direct method, where an elongated simulation cell is sampled through molecular dynamics in the isobaric-isothermal constant area ensemble (NPzzAT). The coarse-grained nature of the force field allows for the accelerated calculation of relative large systems. The interfacial properties of the pure components are described with an error of less than 5% over the whole fluid range. The binary interaction parameters needed for the accurate description of the interfacial properties are obtained from a fit interfacial tensions of the constituent binaries at lower pressures and temperatures and is taken as a constant for all conditions and mixtures studied.

481321 Industrial Fluid Properties Simulation Collective 9th Challenge

Abstract:

This yearâ€™s Industrial Fluid Properties Simulation Collective Challenge is dedicated to high temperature and pressure processing of emulsions. The 9th Challenge aims to test the ability of molecular modeling approaches to predict water/oil interfacial tension (IFT) at elevated temperatures (T) and pressures (P) for three different oils with water, namely n-dodecane, toluene and a 50:50 wt/wt mixture of n-dodecane and toluene. Within the framework of this challenge, we have proceeded to calculate the IFTs of these mixtures by means of molecular dynamics (MD) simulations, the results of which are presented in this report.

474490 Molecular Dynamics Study of Surfactant and Oil Interactions at Saltwater-Air Interfaces

Abstract:

For water treatment purposes, the separation processes involving surfactants and crude oil at seawater-air interfaces are of particular importance for chemical and energy industries. Although there is some experimental work in the area, little progress has been made in fundamentally understanding the nanoscale phenomena of surfactants on oily saltwater-air interfaces. Thus, we focus on using molecular dynamics to simulate the interface with various types of crude oil and surfactants. Preliminary results have shown that with an increasing surface excess of methyl isobutyl carbinol (MIBC) surfactant, there is a decrease in surface tension on an interface composed of the single point charge extended (SPC/E) water model. This is consistent with trends reported in literature and does not contradict the transferable potentials for phase equilibria (TraPPE) force field being utilized. Future work will include investigating other alcohols and glycol ether surfactants as well as light and heavy crude oil types represented by linear alkanes of different molecular weight. An experimental component will be included to assist in verifying model parameters and measurements. By measuring surface tension and characterizing the oil adsorption behavior, the project is aiming to predict surfactant performance for experimental testing under cold seawater conditions.

Molecular-scale Description of SPAN80 Desorption from the Squalane-Water Interface

L. Tan,[1,*] L. R. Pratt,[1,†] and M. I. Chaudhari[2,‡]

[1]*Department of Chemical and Biomolecular Engineering, Tulane University, New Orleans, LA 70118*
[2]*Center for Biological and Material Sciences, Sandia National Laboratories, Albuquerque, NM 87185*
(Dated: April 4, 2016)

Extensive all-atom molecular dynamics calculations on the water-squalane interface for nine different loadings with sorbitan monooleate (SPAN80), at $T = 300$K, are analyzed for the surface tension equation of state, desorption free energy profiles as they depend on loading, and to evaluate escape times for absorbed SPAN80 into the bulk phases. These results suggest that loading only weakly affects accommodation of a SPAN80 molecule by this squalane-water interface. Specifically, the surface tension equation of state is simple through the range of high tension to high loading studied, and the desorption free energy profiles are weakly dependent on loading here. The perpendicular motion of the centroid of the SPAN80 head-group ring is well-described by a diffusional model near the minimum of the desorption free energy profile. Lateral diffusional motion is weakly dependent on loading. Escape times evaluated on the basis of a diffusional model and the desorption free energies are 7×10^{-2} s (into the squalane) and 3×10^{2} h (into the water). The latter value is consistent with irreversible absorption observed by related experimental work.

I. INTRODUCTION

COREXIT 9500 is a standard dispersant used in response to oil spills [1]. Confronting molecular-scale theory, the formulation of COREXIT 9500 is non-trivial and has evolved to address issues identified by decades of experience [2]. Therefore, this material provides context for development of theory, in addition to experiment and simulation, that might provide molecular insight into the structure and function of oil-water-surfactant systems.

COREXIT 9500 includes sorbitan monooleate (SPAN80, FIG. 1), an ethoxylated sorbitan monooleate (TWEEN80), the anionic surfactant sodium dioctyl sulfosuccinate (NaDOS), and alkane solvent (NORPAR 13) [1]. Each of these components, and their proportions, have been chosen to achieve design characteristics [2]. In each case, the statistical mechanical theory that would provide quantitative molecular-scale explanation is unavailable. NaDOS provides one example: the molecular theory of electrolyte effects on the thermodynamics, structure and dynamics of aqueous solution interfaces is a long-standing [3–5] and current research target [6]. As another example, the specific structural and dynamical description of the ethoxylated chains attached to the TWEEN80 head-group, and the contrast to the SPAN80 case, is not available though this is an area of significant recent interest [7–11].

In building the basic molecular theory, it is natural to study the effects arising from the several components separately, and then to study their various interactions. Here we study the solution interface between water and a model oil phase, with SPAN80 at a wide range of loadings. We follow the experimental work of Reichert and Walker [12] in adopting squalane as the model oil phase, though they treated TWEEN80 without SPAN80, and water with non-zero NaCl electrolyte concentration to correspond to seawater. We start with SPAN80 alone because that is simpler, but further complicating features might be added after this initial step [13].

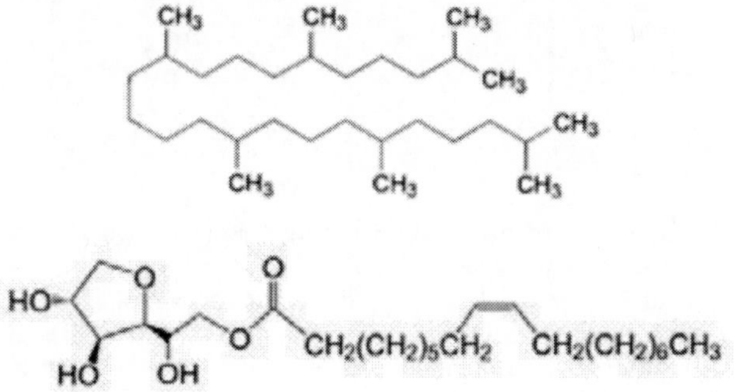

FIG. 1. Chemical structures of squalane (upper) and sorbitan monooleate (SPAN80, lower).

An intriguing aspect of the experimental work [12] was the demonstration of irreversible absorption behavior of TWEEN80 for surface tensions *above* a critical value near 32 mN/m, *i.e.*, for low surface loadings. That behavior was observed to be independent of the ionic strength of the aqueous phase. Operational irreversibility might be traced to rates of desorption, and those rates might be influenced by entanglement of the ethoxylated head-group of the TWEEN80 surfactant. Evaluating desorption rates for SPAN80 at this interface provides a baseline result for considering the TWEEN80 case, and those baseline results are a target of the work which follows below.

The most important step in determining desorption rates is to establish the free energy profile for the process. Here, we obtain free energies of desorption of SPAN80

* ltan2@tulane.edu
† lpratt@tulane.edu
‡ michaud@sandia.gov

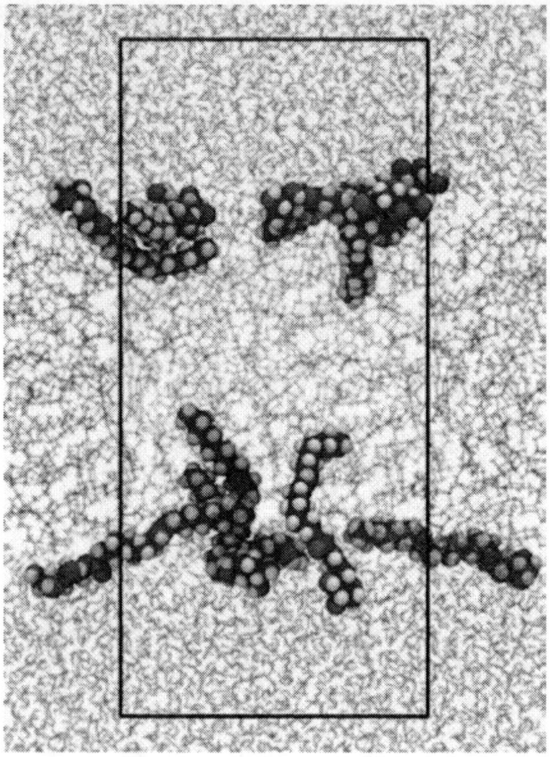

FIG. 2. System for interface simulations. Here the system included 75 squalane molecules (middle layer), 1000 water molecules (top and bottom), and 5 SPAN80 molecules loaded at each interface. Other cases were composed similarly but with different numbers of SPAN80 molecules loaded at the interfaces.

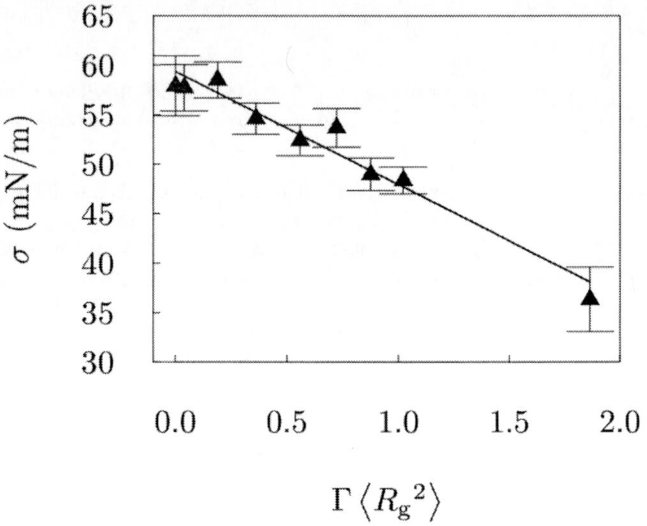

FIG. 3. The surface tension variation demonstrates the expected trend of decreasing σ with increasing surface desorption Γ, here scaled by the mean-square radius of gyration of a SPAN80 molecule in bulk squalane, $\langle R_g^2 \rangle^{1/2} = 0.757$ nm. These results were produced by standard MD simulation as described in Sec. IV. $n_{\mathrm{SPAN80}} = 0$, 1, 5, 10, 15, 20, 25, 30, 50 SPAN80 molecules/interface cases were treated. The statistical uncertainties indicated are 95% confidence intervals obtained by a bootstrap technique based on surface tensions extracted over $\Delta t = 1$ ns segments along 100 ns simulation for each case.

from the water-squalane interface, stratifying the free energy changes with a standard *windowing* approach and exploiting parallel tempering [14] to investigate sampling sufficiency for the individual strata. FIG. 2 shows the case with 5 SPAN80 molecules absorbed at each interface. The windowing not only provides the free energy profile for the desorption but also the net free energy for the transfer of a SPAN80 molecule from water to squalane. Methodological specifics are collected in Sec. IV below, except where they are pertinent to an isolated facet of the results. Preceding simulations studied SPAN80 micelles in the context of drug delivery applications [15].

A distinguishing aspect of our work is that it is that all-atom models are treated exclusively, in contrast to coarse-grained models implemented in CHARMM [16], MARTINI [17] and SDK packages [18]. Important work on escape times of nonionic surfactants from micelles and hydrophobic surfaces has featured coarse-grained MD models [19]. The present work is restricted to planar interfaces.

II. RESULTS AND DISCUSSION

Surface tensions were obtained standardly (Sec. IV) from molecular dynamics simulations for nine surface loadings (FIG. 3). When the absorption Γ vanishes, the surface tension (FIG. 3) agrees well with experiment [12], and the surface tensions decrease with increasing Γ, as expected. Comparing the standard MD results with parallel tempering values (FIG. 4) shows consistent behavior and remarkable simplicity over our whole range of loadings. A maximum loading for SPAN80 naturally should be higher than the maximum loading for TWEEN80, but Γ for our strongly loaded case is about a factor of three (3) higher than the estimated maximum surface coverage for TWEEN80 [12]. A weakly loaded regime is identified for $\Gamma < 0.17/\langle {R_g}^2 \rangle$, about 10% of our highest loading.

Desorption free energy profiles were obtained for the unloaded case ($n_{\mathrm{SPAN80}} = 0$/interface) and our strongly loaded case ($n_{\mathrm{SPAN80}} = 50$/interface). The free energy profiles (FIGs. 5 and 6) are remarkably similar for the two cases. The desorption free energy barrier from interface to squalane phase is around 8 kcal/mol and the free energy barrier is 19.6 kcal/mol from interface to bulk water.

To analyze the kinetics of the desorption process, we assume that the motion of the centroid of a SPAN80

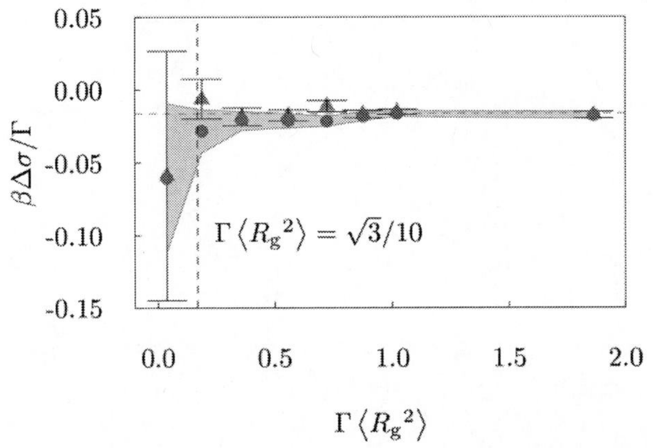

FIG. 4. Dependence of surface stress compressibility factor on surface loading. The ordinate is the surface tension difference $\Delta\sigma \equiv \sigma(\Gamma) - \sigma(\Gamma = 0)$, scaled as indicated with $\beta = 1/k_BT$. The red disks are the data from parallel-tempering simulations as described in Sec. IV. The blue triangles are the data from standard MD simulations. The vertical dashed line indicates $\Gamma\langle R_g{}^2\rangle = \sqrt{3}/10$ as a boundary for dilute loading of this interface. This is based on the idea that a uniform sphere of radius r implies $\langle R_g{}^2\rangle = 3r^2/5$. If those spheres were closely packed $\Gamma = 1/\left(2\sqrt{3}r^2\right)$. The vertical dashed line marks that value. The situation of FIG. 2 corresponds to that low-coverage boundary.

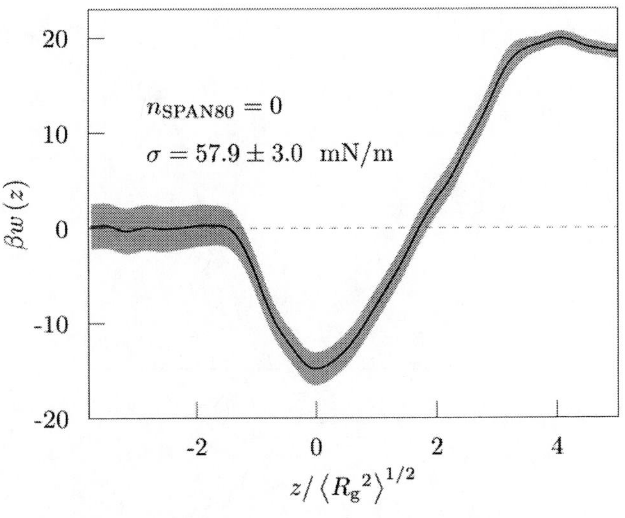

FIG. 5. The potential of the average force on the centroid of a SPAN80 head-group ring in the case of the bare water-squalane interface. $\langle R_g^2\rangle^{1/2} = 0.757$ nm. The standard stratified sampling approach used 150 windows used to cover the whole z range (FIG. 2). Calculations for each window ran for 30 ns. $w(z)$ is reconstructed by the weighted histogram analysis method. The blue band depicts statistical uncertainties of ± 1 standard error estimated pointwise on the basis of a bootstrap resampling of our results. The energy barrier from interface to squalane phase is around 8.2 kcal/mol and the energy barrier from interface to bulk water is 19.6 kcal/mol.

molecule head-group ring may be described by the Smoluchowski equation

$$\frac{\partial}{\partial t}P(z,t|z_0) = D\frac{\partial}{\partial z}\left(\beta w'(z) + \frac{\partial}{\partial z}\right)P(z,t|z_0)\ , \quad (1)$$

with $P(z,t|z_0)$ the conditional probably density for a centroid of a SPAN80 molecule head-group ring to arrive at z after a time t from an initial location z_0 [20]. Motions parallel to the interface, in the x and y directions, are separable in this description. We will use this basis to evaluate the mean first passage time for escape of a SPAN80 molecule from the interface. This description uses $w(z)$ obtained above, but requires also the kinetic parameter D, a self-diffusion coefficient.

To evaluate D, we ran another standard MD simulation preserving sufficient time resolution. Specifically, we extracted a windowed configuration from the strongly loaded interface simulation, then extended the MD simulation for 100 ns with configurations saved every 10 fs. The first 20 ns of this trajectory was discarded as further aging, and we naturally used results for all SPAN80 molecules present. The centroid of a SPAN80 head-group ring is followed with $\delta z = 0$ locating the minimum of $w(z)$ (FIG. 6). The centroid of a SPAN80 molecule head-group ring wanders near the minimum of $w(z)$, and we linearize $w'(z) \approx \kappa\delta z$. The Langevin equation

$$\delta\dot{z}(t) + \beta\kappa D\delta z(t) = R(t)\ , \quad (2)$$

with $R(t)$ the random-force exhaustively discussed elsewhere [21], then corresponds to Eq. (1). Establishing the necessary force constant by $\beta\kappa\langle\delta z^2\rangle = 1$ gives

$$\langle\delta z(0)\delta z(t)\rangle = \langle\delta z^2\rangle\,\mathrm{e}^{-Dt/\langle\delta z^2\rangle}\ . \quad (3)$$

The observed displacement time-correlation-function (FIG. 7) relaxes exponentially, confirming the basic kinetic description. The slight deviation from exponential relaxation (FIG. 7) could be due to the linearization $w'(z) \approx \kappa\delta z$. The self-diffusion coefficient is $1.7 \times 10^{-6}\mathrm{cm}^2/\mathrm{s}$, about a factor of 10 less than a corresponding value for liquid water, and about half of the self-diffusion coefficient of propylene carbonate [22]. Experimental results for diffusion coefficients of nonionic surfactants range from $1\times10^{-6}\mathrm{cm}^2/\mathrm{s}$ to $8\times10^{-6}\mathrm{cm}^2/\mathrm{s}$ [23–26]. The present value corresponds to relaxation times $\langle\delta z^2\rangle/D$ of about 2 ns (FIG. 7). Lateral diffusion is slightly faster here (FIG.8), but slows slightly with increasing interface loading.

With this value of D, we adapt the theory of first passage times [21] to

$$\langle\tau\rangle = \left(\frac{2\pi\langle\delta z^2\rangle}{D}\right)\int_0^{\delta z^\ddagger}\mathrm{e}^{\beta\Delta w(z)}\frac{\mathrm{d}\delta z}{\sqrt{2\pi\langle\delta z^2\rangle}} \quad (4)$$

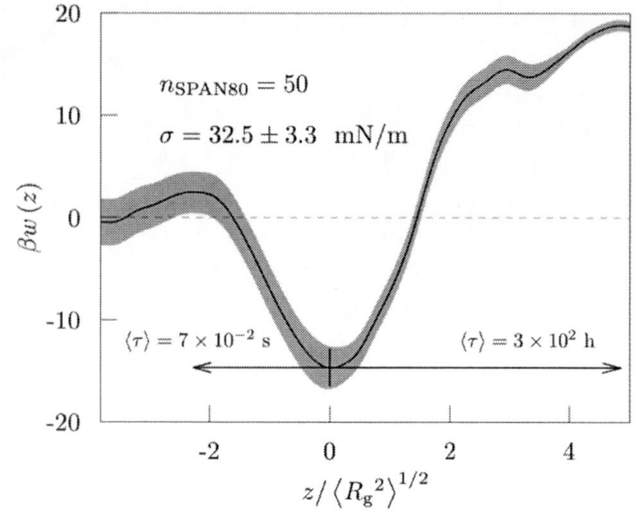

FIG. 6. The potential of mean force for the strongly loaded water-squalane interface. The abscissa is the z coordinate of the centroid of a SPAN80 molecule head-group ring scaled by mean-square radius of gyration of a SPAN80 molecule in bulk squalane, $\langle R_{\rm g}^2\rangle^{1/2} = 0.757$ nm. The standard stratified sampling approach used 190 windows to cover the whole z range. Calculations for each window ran for 60 ns. $w(z)$ is recomposed by the weighted histogram analysis method. The blue band depicts statistical uncertainties of ± 1 standard error estimated pointwise on the basis of a bootstrap resampling of our results. These results show an energy barrier from interface to squalane phase is around 8.7 kcal/mol and the energy barrier from interface to bulk water is around 19.6 kcal/mol.

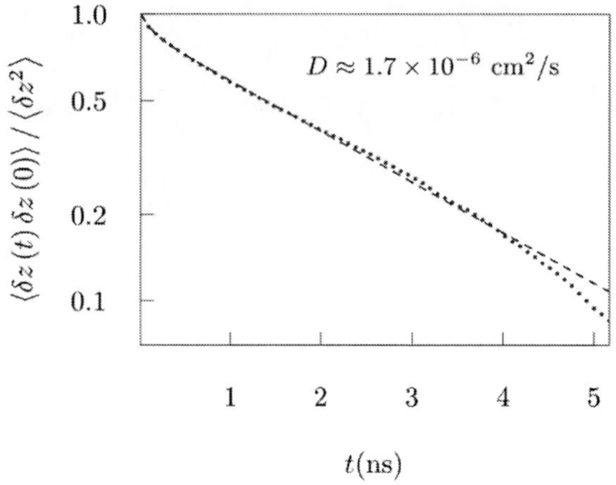

FIG. 7. The autocorrelation function for displacement of the centroid of a SPAN80 head-group ring at the strongly loaded interface. The black dashed line is the function $A\mathrm{e}^{-t/\tau_1} + (1-A)\mathrm{e}^{-t/\tau_2}$ fit to the primitive results (blue dots). $A = 0.12$, $\tau_1 = 0.12$ ns, and $\tau_2 = 2.46$ ns.

where $\Delta w(z) = w(z) - w_{\rm min}$, and $\delta z^\ddagger$ is the displacement from the minimum at $\delta z = 0$ to the barrier configuration. We then find (FIG. 6) the mean first passage time of 0.07 s, from the interface to the squalane phase, while the mean first passage time from the interface to bulk water is about 3×10^2 h. Differences in the barrier shapes in the two directions play a role in determining that value, since the ratio of these times is about a factor of four less than the naive ratio of $\mathrm{e}^{\beta\Delta w^\ddagger} \approx \mathrm{e}^{18}$.

III. CONCLUSIONS

These results suggest that loading only weakly affects accommodation of a SPAN80 molecule by this squalane-water interface. Specifically, the surface tension equation of state (FIG. 4) is simple through the range of high tension to high loading studied, and the desorption free energy profiles are weakly dependent on loading here (FIGs. 5 and 6). The free energy of transfer of a SPAN80 molecule from water to squalane is about -28 kcal/mol. The perpendicular motion of the centroid of the SPAN80 head-group ring is well-described by a diffusional model near the minimum of the desorption free energy profile (FIG. 7). Lateral diffusional motion is weakly dependent on loading (FIG. 8). Escape times approximated on the basis of a diffusional model and the desorption free energies are 7×10^{-2} s (into the squalane) and 3×10^2 h (into the water), the latter value consistent with irreversible absorption observed by recent experimental work [12] on a related system.

IV. METHODOLOGICAL SPECIFICS

The GROMACS [27] 4.6.7 molecular dynamic simulation package was used for all calculations. The chain molecules were represented by OPLS-AA force field [28] and the SPC/E model [29] was chosen for water. The Nose-Hoover thermostat [30] maintained the constant temperature and Parrinello-Rahman barostat [31] kept pressure at 1.0 atm. Long-range electrostatic interactions were treated in standard periodic boundary conditions using the particle mesh Ewald method with a cutoff of 1.0 nm. The chemical bonds involving hydrogen atoms were constrained by the LINCS algorithm [32].

In evaluating surface tensions, we applied standard MD simulation and checked the sampling sufficiency with parallel tempering calculations. Nine different surface loadings were investigated, with $n_{\rm SPAN80} = 1, 5, 10, 15, 20, 25, 30$ per interface. The squalane phase included 75 squalane molecules and was bounded by a water phase, bottom and top layers, of 1000 water molecules. For the highest loaded ($n_{\rm SPAN80} =$ 50/interface) and unloaded ($n_{\rm SPAN80} =$ 0/interface) cases, a larger system of 100 squalane molecules and 3000 water molecules was simulated. We expanded the

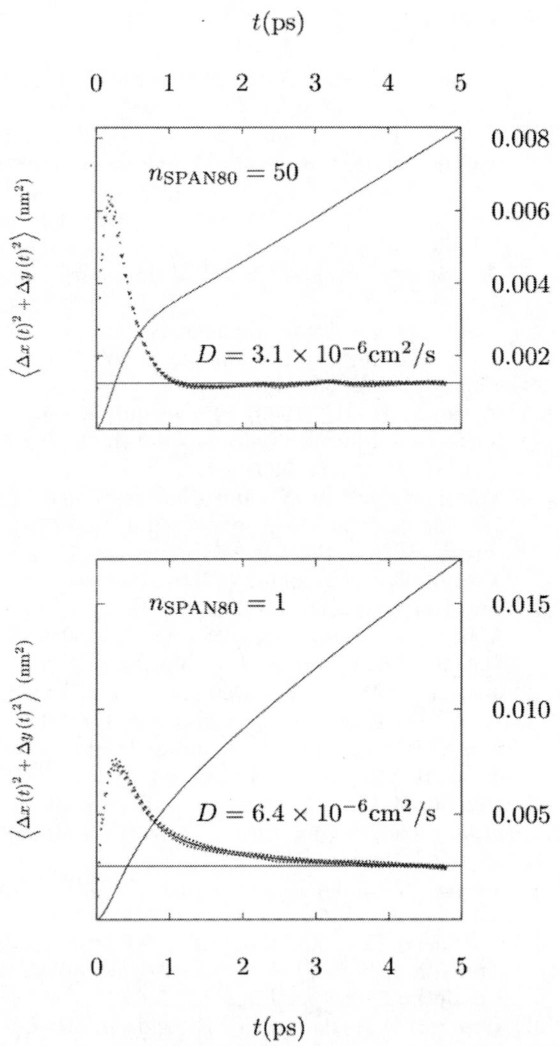

FIG. 8. Lateral diffusive motion of the SPAN80 ring centroid and its dependence on loading. The plotted blue marks are estimates of the time derivative of the solid lines. Lateral diffusion coefficients are slightly larger than the perpendicular coefficient (FIG. 7), but interfacial crowding has the expected effect of slowing the diffusion slightly.

system for those two cases to accommodate a consistent comparison for desorption free energy evaluations. Our standard procedure carried-out an energy minimization calculation and density equilibration followed by a 100 ns production run with constant particle number, pressure, and temperature (NPT) conditions. The parallel tempering calculations used 48 replicas spanning the 260-450K temperature range. Trajectories ran for 50 ns and the resulting exchange probability between neighboring temperatures was around 20%.

We utilized the windows sampling approach to evaluate the desorption free energy profile. To generate initial configurations for each window, we placed one more SPAN80 in the water phase at the position where $z = 1.0$ nm. A pulling force was applied to the centroid of the SPAN80 head-group ring to pull it across the whole system. To achieve high resolution for $w(z)$, 150 windows were used to cover the whole z range in unloaded system while 190 windows are utilized to cover the whole z range in strongly loaded system. Due to the different complexity between two cases, the extent of the MD trajectory/window differs. Trajectories ran for 30 ns/window in unloaded case while the calculations extended to 60 ns/window in the strongly loaded case.

V. ACKNOWLEDGEMENT

Sandia National Laboratories is a multi-program laboratory managed and operated by Sandia Corporation, a wholly owned subsidiary of Lockheed Martin Corporation, for the U.S. Department of Energy's National Nuclear Security Administration under contract DE-AC04-94AL85000. The financial support of Sandia's LDRD program and the Gulf of Mexico Research Initiative (Consortium for Ocean Leadership Grant SA 12-05/GoMRI-002) is gratefully acknowledged.

[1] *Understanding Oil Spill Dispersants: Efficacy and Effects*; National Academies Press, Washington DC, 2005.

[2] Riehm, D. A., and McCormick, A. V. (2014) The Role of Dispersants Dynamic Interfacial Tension in Effective Crude Oil Spill Dispersion. *Marine pollution bulletin 84*, 155–163.

[3] Nichols, A. L., and Pratt, L. R. (1982) Disentanglement of Hydrophobic and Electrostatic Contributions to the Film Pressures of Ionic Surfactants. *Faraday Symp. Chem. Soc. 17*, 129–140.

[4] Wilson, M. A., Nichols, A. L., and Pratt, L. R. (1984) Hydrophobic Interaction of Amphiphilic Ions with Water–Hydrocarbon Liquid Interfaces. *J. Chem. Phys. 81*, 579–580.

[5] Nichols, A. L., and Pratt, L. R. (1984) Salt Effects on the Surface Tensions of Dilute Electrolyte Solutions: The Influence of Nonzero Relative Solubility of the Salt between the Coexisting Phases. *J. Chem. Phys. 80*, 6225–6233.

[6] Jungwirth, P., and Cremer, P. S. (2014) Beyond Hofmeister. *Nature Chem. 6*, 261–263.

[7] Alessi, M., Norman, A., Knowlton, S., Ho, D., and Greer, S. (2005) Helical and Coil Conformations of Poly(ethylene glycol) in Isobutyric Acid and Water. *Macromol. 38*, 9333–9340.

[8] Norman, A. I., Fei, Y., Ho, D. L., and Greer, S. C. (2007) Folding and Unfolding of Polymer Helices in Solution.

Macromol. 40, 2559–2567.

[9] Norman, A. I., Ho, D. L., and Greer, S. C. (2007) Partitioning, Fractionation, and Conformations of Star Poly(ethylene glycol) in Isobutyric Acid and Water. *Macromol. 40*, 9628–9639.

[10] Chaudhari, M. I., Pratt, L. R., and Paulaitis, M. E. (2014) Concentration Dependence of the Flory-Huggins Interaction Parameter in Aqueous Solutions of Capped PEO Chains. *J. Chem. Phys. 141*, 244908–5.

[11] Chaudhari, M. I., Pratt, L. R., and Paulaitis, M. E. (2015) Loop-Closure and Gaussian Models of Collective Structural Characteristics of Capped PEO Oligomers in Water. *J. Phys. Chem. B 119*, 8863–8867.

[12] Reichert, M. D., and Walker, L. M. (2013) Interfacial Tension Dynamics, Interfacial Mechanics, and Response to Rapid Dilution of Bulk Surfactant of a Model Oil-Water-Dispersant System. *Langmuir 29*, 1857–1867.

[13] Kirby, S. M., Anna, S. L., and Walker, L. M. (2015) Sequential Adsorption of an Irreversibly Adsorbed Nonionic Surfactant and an Anionic Surfactant at an Oil/Aqueous Interface. *Langmuir 31*, 4063–4071.

[14] Earl, D. J., and Deem, M. W. (2005) Parallel Tempering: Theory, Applications, and New Perspectives. *Phys. Chem. Chem. Phys. 7*, 3910–3916.

[15] Han, S. (2013) Molecular Dynamics Simulation of Sorbitan Monooleate Bilayers. *Bull. Kor. Chem. Soc. 34*, 946–948.

[16] Bjelkmar, P., Larsson, P., Cuendet, M. A., Hess, B., and Lindahl, E. (2010) Implementation of the CHARMM Force Field in GROMACS: Analysis of Protein Stability Effects from Correction Maps, Virtual Interaction Sites, and Water Models. *J. Chem. Theory Comp. 6*, 459–466.

[17] Marrink, S. J., Risselada, H. J., Yefimov, S., Tieleman, D. P., and Devries, A. H. (2007) The MARTINI Force Field: Coarse Grained Model for Biomolecular Simulations. *J. Phys. Chem. B 111*, 7812–7824.

[18] Shinoda, W., DeVane, R., and Klein, M. L. (2008) Coarse-grained Molecular Modeling of Non-ionic Surfactant Self-assembly. *Soft Matter 4*, 2454–2462.

[19] Yuan, F., Wang, S., and Larson, R. G. (2015) Potentials of Mean Force and Escape Times of Surfactants from Micelles and Hydrophobic Surfaces Using Molecular Dynamics Simulations. *Langmuir 31*, 1336–1343.

[20] van Kampen, N. G. *STOCHASTIC PROCESSES IN PHYSICS AND CHEMISTRY*; North Holland, 1992.

[21] Zwanzig, R. *NONEQUILIBRIUM STATISTICAL MECHANICS*; Oxford University Press, 2001.

[22] You, X., Pratt, L., and Rick, S. (2014) The Role of Attractive Interactions in the Dynamics of Molecules in Liquids. *arXiv preprint arXiv:1411.1773*

[23] Luschtinetz, F., and Dosche, C. (2009) Determination of Micelle Diffusion Coefficients with Fluorescence Correlation Spectroscopy (FCS). *J. Coll. Interf. Sci. 338*, 312–315.

[24] Miller, R., and Lunkenheimer, K. (1986) Adsorption Kinetics Measurements of Some Nonionic Surfactants. *Coll. Poly. Sci. 264*, 357–361.

[25] Paradies, H. H. (1980) Shape and Size of a Nonionic Surfactant Micelle. Triton X-100 in Aqueous Solution. *J. Phys. Chem 84*, 599–607.

[26] Moorkanikkara, S. N., and Blankschtein, D. (2009) New Methodology to Determine Equilibrium Surfactant Adsorption Properties from Experimental Dynamic Surface Tension Data. *Langmuir 25*, 6191–6202.

[27] Van Der Spoel, D., Lindahl, E., Hess, B., Groenhof, G., Mark, A. E., and Berendsen, H. J. (2005) GROMACS: fast, flexible, and free. *J. Comp. Chem. 26*, 1701–1718.

[28] Jorgensen, W. L., Maxwell, D. S., and Tirado-Rives, J. (1996) Development and Testing of the OPLS All-Atom Force Field on Conformational Energetics and Properties of Organic Liquids. *J. Am. Chem. Soc. 118*, 11225–11236.

[29] Berendsen, H. J. C., Grigerat, J. R., and Straatsma, T. P. (1987) The Missing Term in Effective Pair Potentials. *J. Phys. Chem 91*, 6269–6271.

[30] Evans, D. J., and Holian, B. L. (1985) The Nose-Hoover Thermostat. *J. Chem. Phys. 83*, 4069–4074.

[31] Parrinello, M., and Rahman, A. (1981) Polymorphic Transitions in Single Crystals: A New Molecular Dynamics Method. *J. App. Phys. 52*, 7182.

[32] Hess, B., Bekker, H., Berendsen, H. J. C., and Fraaije, J. G. E. M. (1997) LINCS: A Linear Constraint Solver for Molecular Simulations. *J. Comp. Chem. 18*, 1463–1472.

466193 Some Effects of Surface Heterogeneity on the Morphology of Surfactant Self-Assembled Aggregates

Abstract:

Surfactants are ubiquitous chemicals in academia, industry, and consumer products. Just in the oil and gas industry, these compounds can be used for enhanced oil recovery and environmental remediation strategies. In consumer-care products, surfactants find wide use in soaps, shampoos, and many other products. Fundamental investigations have been focused on surfactants for many years, not only because self-assembling systems present a number of questions that are intrinsically interesting, but also for the expected positive impacts in a variety of industrial sectors, including nanotechnology and drug delivery. However, most studies have concentrated on homogeneous systems, either in the bulk, or at contact with pristine surfaces. A number of recent discoveries suggest that perhaps the behavior of surfactants on heterogeneous surfaces differs compared to that expected based on results obtained on homogeneous substrates. In this presentation we will present recent results obtained using multiple computational strategies to unveil the effects of heterogeneous surface properties on the morphology of self-assembled surfactant aggregates. We will also discuss how our results can be explained by an expression for the system free energy that accounts for the elastic deformation of the self-assembled surfactant film.

465537 Adsorption of Surfactants on Metallic Surfaces Studied Using Molecular Simulations

Abstract:

Surfactant molecules, such as hydrocarbons with imidazoline derivative polar group, are popular in the oil and gas industry as corrosion inhibitors. Corrosion inhibitors retard corrosion by physisorbing on metallic surfaces. Experimental observations suggest that, upon adsorption to metallic surfaces, these molecules form a self-assembled monolayer (SAM). While several such molecules are extensively used in the industry, there is a need to optimize the performance because: (a) adsorbed inhibitor films do not completely eliminate corrosion; and (b) their efficacy depends on operating conditions, such as solution pH, temperature, oil composition, etc. The search for new corrosion inhibitors is largely serendipitous as the mechanism of their action is poorly understood. In this talk, molecular simulation results of adsorption characteristics and formation of SAM of these molecules as function of their concentration and chemistry will be discussed. These results will be useful for experimentalists seeking to design novel molecules as efficient corrosion inhibitors for specific applications.

458366 Quantum Chemistry Study of Curvature Effects on Boron Nitride Nanotube/Nanosheet for Gas Adsorption

Abstract:

Boron Nitride (BN) layered nanomaterials have large specific surface area and excellent thermal and structural stabilities, which make them as promising materials for gas adsorption applications. However, BN structures have not been widely studied to date. How the BN surface structures can affect the gas adsorption property and further potential to industrial application is still unknown. In the present study, quantum chemistry calculations were performed to investigate the effect of the surface curvature of a BN nanotube/nanosheet on gas adsorption. Curved Boron Nitride layers with different curvatures interacting with a number of different gases including noble gases, oxygen, and water on both their convex and concave sides of the surface were studied by density functional theory (DFT) with a high level dispersion corrected functional. Potential energy surfaces of the gas molecules interacting with the selected BN surfaces were investigated. In addition, the charge distribution and electrostatic potential contour of the selected BN surfaces are discussed. The results reveal how the curvature of the BN surfaces affects gas adsorption. In particular, small curvatures lead to slight difference in the physisorption energy, while large curvatures present distinct potential energy surfaces, especially for the short-range repulsion. The present study suggests that the curvature effect should be considered in the design of BN layered nanomaterials for gas adsorption and storage applications.

461425 Molecular Simulation of Metal-Ionic Liquid Interfaces

Abstract:

Surface interactions between metals and room temperature ionic liquids (RTILs) give rise to promising applications in catalysis, tribology, electrochemical deposition, and energy storage. Despite this, nanoscale features of the electric double layer (EDL) are poorly understood for more than a small number of metals. Here, we use a combination of molecular dynamics (MD) simulation, atomic force microscopy (AFM), and neutron reflectometry (NR) to study the structural ordering of ionic liquids at the interface of metal surfaces. Recent advances in AFM and NR have enabled the detection of ion layering at the EDL and structural responses to applied electrical potential. These experimental probes enable comparison to ion density profiles derived from MD. We initially consider the RTIL EMIM-Tf2N on a molybdenum surface. Due to a lack of appropriately generalizable force fields for a pure Mo surface, we parameterized Lennard-Jones interactions using contact angle experiments as target data. The derived fluid-solid interactions are used to explore the structure and orientation of RTIL molecules at the metal interface. Systems with and without applied electrical potentials are explored. The generalizability of this method to other surfaces, including bismuth, tin, and carbon, is explored.

472171 Effects of Fluoroethylene Carbonate and Dissolved Manganese on the ANODE SEI Layer Using Reaxff-Based Molecular Dynamics

Abstract:

Lithium manganese oxide cathode and graphite anode based batteries are sought after for high power applications like electric vehicles and energy storage applications, due to their high capacity, low cost, and low toxicity. There are however well-documented problems with capacity fade and cycle life of this battery chemistry, caused by the dissolution of divalent manganese from the cathode surface and deposition of dissolved manganese onto the anode. Most commercial cells use electrolyte additives to develop a stable solid-electrolyte interface layer (SEI) on the surface of the anode to increase cycle life. The speciation and reaction mechanisms of these electrolyte additives are nonetheless not well understood.

In this study, reactive force field-based molecular dynamics (ReaxFF MD) was applied to investigate the decomposition of fluoroethylene carbonate (FEC), a commonly used electrolyte additive, on the surface of a graphite anode. The ReaxFF MD simulations indictate that, contrary to widely-held belief, lithium carbonate, not lithium fluoride, is the product of FEC decomposition. The fate and effect of dissolved manganese ions on the anode SEI layer was also studied. Here, the ReaxFF MD simulations reveal that manganese at the anode-electrolyte interface facilitates the polymerization of ethylene carbonate, forming polymer chains that extend from the anode surface into the electrolyte.

469199 Formation Mechanisms for Microscopic Precursors in Droplet Spreading

Abstract:

Molecular precursor are ultrathin films that preced spreading droplets. While the mass transport mechanisms and dynamics of molecular precursors are known, what is necessary for precursors to form, and the reason why different kinds of precursors exist, remain unclear. Based on large-scale molecular dynamics simulations of a relatively simple molecular model, we demonstrate that as the interaction between the droplet and substrate becomes more favorably attractive, the spreading process passes through successive regimes with no precursor, a monolayer precursor, and a continuously growing precursor, respectively. We establish general guidelines governing both the nature of the spreading process as well as the shape of the precursor, when it exists.

474750 Mathematics for Data-Driven Modeling - the Science of Crystal Balls

Abstract:

In mathematical modeling one typically progresses from observations of the world (and some serious thinking!) to equations for a model, and then to the analysis of the model to make predictions. Good mathematical models give good predictions (and inaccurate ones do not) - but the computational tools for analyzing them are the same: algorithms that are typically based on closed form equations. While the skeleton of the process remains the same, today we witness the development of mathematical techniques that operate directly on observations -data-, and "circumvent" the serious thinking that goes into selecting variables and parameters and writing equations. The process then may appear to the user a little like making predictions by "looking into a crystal ball". Yet the "serious thinking" is still there and uses the same -and some new- mathematics: it goes into building algorithms that "jump directly" from data to the analysis of the model (which is never available in closed form) so as to make predictions. I will present a couple of efforts that illustrate this
new path from data to predictions. It really is the same old path, but it is travelled by new means.

461842 Decoding Common Features of Protein-Nanoparticle Interactions

Abstract:

Nanoparticles have been recognized as key to revolutionizing the healthcare industry. However, excitement about the applications envisioned for nanoparticles is tempered by concerns about their potential toxicity. Both healthcare applications and potential toxicity require knowledge of nanoparticleâ€™s interactions with proteins. Our goal is to discover the mechanisms that govern protein-nanoparticle interactions and to develop a tool that can predict interaction strengths based on knowledge solely of the proteinâ€™s sequence and the nanoparticleâ€™s chemicophysical properties. We calculate the interaction energies between 2315 proteins and three bare gold nanoparticles (diameter = 1.0, 2.0 and 4.0 nm) and identify the binding sites using computer simulations. This allows us to rank the protein-nanoparticle interaction strengths, explore the chemicophyiscal properties of the binding sites, and develop models that can predict nanoparticle-protein interaction energies. Our results show that glutamic acid and lysine have the highest chance to appear on nanoparticle binding sites. We develop several models that can qualitatively predict the interaction energy.

471936 Design of Optimal Experimental Probes for Protein Dynamics Using Machine Learning and Variational Approach to Modeling Conformational Kinetics

Abstract:

In this talk, we plan to answer a broad question about design of experiments and validation of protein conformational ensembles, what is the minimum experimental information required for describing the conformational dynamics of a protein? Markov state models of protein dynamics describe the transitions between the conformational states of proteins using a transition probability matrix built from the raw simulation data. The eigenvalues and eigenvectors/eigenfunctions of the transition matrix provide the timescales and the population shift between states due to a particular dynamic process respectively. Considering only the m-slowest processes, an ideal set of experimental observations would unequivocally describe these m-slowest processes using only m-experimental observables. Our machine learning protocol involves using the variational approach for building Markov state models that provides approximation of the first m eigenfunctions of the molecular dynamics propagator using all possible experimental probe positions on the protein and picks the optimal set of probe locations that best describe the m-slowest processes. In this work, we combine two central ideas from machine learning and chemical physics - hyperparameter selection via cross-validation and variational approaches for linear operator eigenproblems; to create a new method for predicting observables that optimally discriminate between alternative models for molecular kinetics constructed from MD simulations. This protocol could also be used with a broad range of experimental techniques or a combination of experimental techniques. We present results from application of this protocol to three systems of biological importance, conformational change in membrane peptide transporter $PepT_{SO}$ and Î²2-Adrenergic receptor, and folding of WW domain.

469029 Guiding Experiments Towards New Functional Materials with Informatics

Abstract:

In this talk, we will discuss our informatics approach that integrates machine learning, symmetry analysis and density functional theory (DFT) to guide experiments in search of novel $BaTiO_3$-based ferroelectric materials with high Curie temperature (T_C). $BaTiO_3$-based materials have attracted significant attention as a potential alternative for Pb-free piezoelectrics. In the development of $BaTiO_3$-based piezoelectrics, one of the common strategies involves mixing two end member compositions that have Cubic (*Pm-3m*) to Tetragonal (*P4mm*) and Cubic to Rhombohedral (*R3m*) transformations. Our design objective is to identify suitable chemical substitutions at the Ba- and Ti-sites such that the T_C of the end members are high. We perform DFT calculations on undoped $ATiO_3$ and $BaBO_3$ perovskites in *P4mm* and *R3m* symmetries, respectively; A and B are divalent and tetravalent cations, respectively. From symmetry-mode analysis, we decompose the two ferroelectric phases (*P4mm* and *R3m*) in terms of irreducible representations (irreps). These irreps, along with the unit cell volume from DFT, serve as *feature sets* for understanding $ATiO_3$ and $BaBO_3$ crystal chemistries and their role in impacting the ferroelectric distortions. We then link these features with known experimental T_C data using machine learning methods and predict a new $CdTiO_3$-$BaTiO_3$ solid solution for experimental synthesis. We performed experiments and found that our predictions do not agree with the measurements. The reason for the discrepancy is attributed to the anomalous behavior of the $CdTiO_3$-$BaTiO_3$ solid solution, the phase diagram of which is found to be different from any of the $BaTiO_3$-based solid solutions studied so far.

472588 Pushing the Frontiers of Atomistic Modeling Towards Predictive Design of Materials

Abstract:

The ability to perform accurate calculations efficiently is crucial for computational materials design. In this talk, we will discuss our stream-lined approach to force field development using first principles density functional theory training data and machine learning algorithms. Our objective has been to develop new, first-principles based, more accurate and more robust inter-atomic potentials for accurate simulations of dynamical processes at reactive interfaces and low dimensional systems such as clusters and molecules. The procedure involves several steps including (a) generation and manipulation of extensive fitting data sets through electronic structure calculations, (b) defining functional forms, (c) formulating novel highly optimized fitting procedures, (d) dual-Hamiltonian optimization to leverage classical FFs with more accurate approaches, and (d) subsequently coding and implementing these algorithms on high performance computers (HPCs). We will also discuss the validation of this approach on several diverse material systems ranging from precious metal nanocatalysts to newly discovered two dimensional materials such as stanene and silicene.

469177 Design of Ternary Transparent Conducting Oxides

Abstract:

Transparent conducting oxides (TCOs) are well-developed and commercialized class of wide-bandgap semiconductors that are crucial for the function of many electronic devices. Recent experimental work has demonstrated bandgap engineering in ternary $(Al_xGa_yIn_{1-x-y})_2O_3$ over several ~ 3 eV by adjusting the ratio of In/Ga[1] and Ga/Al.[2]

The focus of this work is an examination of the phase diagram for ternary $(Al_xGa_yIn_{1-x-y})_2O_3$ materials using DFT-based cluster expansion models combined with fast stochastic optimization techniques (*e.g.*, nested sampling). This combined computational approach allows for both an efficient search of the stable and metastable configurations for $(Al_xGa_yIn_{1-x-y})_2O_3$ at various lattice types and the consideration of entropy on the relative stability of ternary TCOs. Statistical learning, in particular compressed sensing, is used to efficiently identify a structure-property relationship between the targeted properties (*e.g.*, mobilities and optical transparency) and the fundamental chemical and physical parameters that control these properties.

[1] Zhang et al., *Solid State Commun*, 186, 28 (2014).

[2] Ito et al., *Jpn. J. Appl. Phys.*, 51, 100207 (2012); Zhang et al., *Appl. Phys. Lett.*, 105, 162107 (2014).

472938 Development of Empirical Charge Transfer Interatomic Potential for Tantalum Oxide Nanostructures from First Principle Calculations

Abstract:

There has been considerable interest over the years in metal oxides and their nanostructures owing to their exotic properties making them ideal candidates for several applications, notably in electronics, thermal management, catalysis and biochemistry. In particular, mesoporous (2-50 nm pore sizes) tantalum oxide nanostructures can be used for photocatalytic decomposition of water, thermal insulation and molecular adsorption, while Ta_2O_5nanoparticles have applications in bio-imaging due to their inertness and non-toxicity. It is essential to obtain a clear fundamental understanding of the structure and chemistry of metal oxide nanostructures to optimize the material performance in these applications.

Atomistic molecular dynamics (MD) simulations provide an ideal tool to investigate the underlying mechanisms at play in the nanoscale processes around these nanostructures and to decode the structure-property relationship. The key to a reliable MD simulation is the underlying empirical force field that describes the interatomic interactions. While robust interatomic potentials have been developed for several metal oxides, namely, alumina, titanium oxide, iridium oxide etc., there is no available empirical charge transfer force field to simulate the behavior of tantalum oxide nanostructures and metal-oxide interfaces. This is primarily due to a lack of comprehensive structural and energetics information, both experimental and ab initio theory, on the different polymorphs of Ta_2O_5.

In this talk, we present a detailed tabulation of ab initio density functional theory (DFT) calculations of structure, cohesive energies and elastic properties for the three main experimentally observed polymorphs of Ta_2O_5, namely, hexagonal, orthorhombic and monoclinic. The DFT-generated dataset is used to parameterize an EAM+QEQ (Streitz-Mintmire) charge transfer interatomic potential for tantalum oxide via a genetic algorithm based framework.

Streitz-Mintmire force field is a variable charge potential, which can account for dynamic charge transfer, and has been demonstrated to be a reasonable potential for single-metal oxides. In particular, oxidation of metallic nanoparticles can be investigated using this charge transfer formalism. The developed force field can also be used to investigate Ta/TaO interfaces, the properties of which are significantly affected by the nanostructure of the interface.

In this work, the developed interatomic potential is used to investigate the oxidation of tantalum thin films and nanoparticles. The thermal and mechanical stability of tantalum oxide nanoparticles and mesoporous nanostructures, and the atomic scale processes (oxidation and sintering) that occur in the vicinity of these nanoclusters are also investigated.

455131 Machine Learning for Advancing Discovery of Novel Thermoelectric Materials. the Thermoel

Abstract:

Based on current estimates 59% of all primary energy produced in the United States is lost as waste heat, recovery of even 1% of this heat by a novel thermoelectric materials would amount to nine times the energy currently produced by photovoltaics. Thermoelectric materials are materials that exhibit a strong coupling between electric and thermal responses, producing an electric potential in the presence of a temperature gradient and vice versa.
The physics governing the material systems is described as a network of models and functional relationships between different scales and modalities (i.e. processing/structure/properties). However, the disjointed nature of materials property databases and relative immaturity of the thermoelectrics multi-scale functional relationships make integrated optimization across numerous length and time scales particularly difficult in this case. A data-driven design of advanced thermoelectrics is possible with tools capable of integrating materials databases into a consistent framework and finding the multi-scale underlying processing/structure/property functional relationships responsible for thermoelectrics performance.
The ThermoEl could be an effective solution to discovery of novel thermoelectrics. It is a predictive toolkit developed based on a supervised machine learning algorithms and known experimental measurements of various materials. Aside from other tools on the market, ThermoEl is based on regression models that predict numerical value of such parameters as Seebeck coefficient for non-stoichiometric materials and Bulk modulus of crystalline compounds. Based on our experience, we would like to discuss tricks and challenges in using data mining approaches with materials databases.

468062 Identifying Descriptors for Dielectric Breakdown Strength Using Genetic Programming

Abstract:

The development of phenomenological theories by identifying relevant descriptors for material properties enables rapid screening large numbers of materials and facilitates the design of new materials. One of the leading challenges for computational researchers is determining the best ways to analyze large material data sets to identify the best descriptors for a given property. In this presentation, we demonstrate the use of genetic programming to identify relevant descriptors of dielectric breakdown based on 82 representative materials. We identified band gap E_g and phonon cut-off frequency w_{max} as the two most correlated features, and new classes of phenomenological models featuring exponential functions of E_g*w_{max} were uncovered. The genetic programming model was found to outperform other models for descriptor identification, and we discuss some of the advantages of the genetic programming approach.

462881 Machine Learning with Structural Fingerprints of Local Particle Environments

Abstract:

As the computational power available to scientists increases, researchers themselves — rather than hardware or algorithms — become the bottleneck in scientific discovery. The computational study of colloidal self-assembly is one area that keenly feels this effect: even after computers generate massive amounts of raw data, performing an exhaustive search to determine what, if any, ordered structures occur in a large parameter space of many simulations can be a long, manual process. Here we demonstrate how "off-the-shelf" machine learning algorithms can be applied to results of self-assembly studies both to find interesting regions in a phase diagram and identify characteristic local environments in simulations in an automated, high-throughput manner for both simple and complex crystal structures. These methods can form the foundation of an intelligent exploration of parameter space — a key component in the process of creating new materials by design.

454766 Using Semi-Supervised Machine Learning to Map the Phase Diagrams of Open Materials Data Sets

Abstract:

The Materials Genome Initiative (MGI) is a government program intended to expedite materials discovery, optimization, and deployment by combining theory, experiments, and data science in an integrated workflow. This effort requires that large quantities of high quality experimental data be made publicly available via minable materials databases for the purpose of validating new data mining techniques and validating theoretical results. High-throughput experimentation has been identified as an important technology for creating large-scale experimental databases that delineate the impact of synthesis, processing and composition on crystal phase stability and figure of merit. To date, however, relatively few of these rich data sets are freely available and those that are available are primarily restricted to unlabeled as-obtained structural data.

Such data is critically important to identifying the materials genome, as it creates the linkage between composition, structure and property. Algorithmic approaches to automated phase diagram mapping have been a hot issue in the high-throughput field for a number of years and most studies have focused on an open FeGaPd data set that has been available for about 10 years. Here, we demonstrate a semi-supervised machine learning technique, SS-AutoPhase, which uses a two-step approach to automatically identify phases within structural data sets. In the first step, clustering analysis is used to automatically select a representative sub-set of samples to be manually analyzed by a human expert. In the second step, these labeled samples are used by an AdaBoost classifier to identify the presence of the different phases in the FeGaPd diffraction data. SS-AutoPhase was used to identify the metallographic phases in 278 diffraction patterns from a FeGaPd sputtered composition spread sample. The accuracy of SS-AutoPhase was greater than 82.6% for all phases when 15% of the diffraction patterns were used for training. Furthermore, the predicted phase diagram of SS-AutoPhase was determined and compared to phase labels from a human expert and other algorithmic approaches. This comparison showed that not only did SS-AutoPhase have very high agreement with the expert phase labels, but

that it was able to determines and correctly identify a previously unreported phase. Finally, I will report on a first-of-its-kind identification of a novel ferromagnetic shape memory alloy via the data mining of an open materials database.

474935 The Prediction of Environmentally Important Properties for Hazardous Chemicals

Abstract:

In this talk the successes in prediction of environmentally important properties of hazardous chemicals using modern computational methods will be viewed. Some of the properties considered are solubility in water, Henryâ€™s law constants, octanol-water partition coefficients, and soil-water partition coefficients. The species include industrial chemicals, explosives and chemical warfare agents.

454967 Corresponding States Theory for Property Prediction at the Nanoscale

Abstract:

The Principle of Corresponding States was first proposed by van der Waals in 1881, based on his equation of state. The Principle made it possible to predict the vapor pressure curve, including the critical and normal boiling points, of liquids for which no measurements exist, based on experimental data for other liquids. It quickly found use in this way for predicting liquid properties for hydrogen and helium, neither of which had been liquefied at the time. When these light gases were later liquefied the predictions were found to be accurate. A more powerful derivation of the Principle is based on statistical thermodynamics, which makes clear the assumptions involved and provides a path to improve the theory so that it is more generally applicable. Subsequent generalizations of the theory of corresponding states, in particular that due to Pitzer, provide the basis for the prediction of many physical properties of homogeneous gases and liquids, and are widely used in process simulation. However, these correlations are not applicable to highly inhomogeneous or nanoscale systems, such as small drops, nanoparticles, adsorbed thin films on solid substrates, polymer brushes, etc.

A statistical mechanical analysis of the properties of a highly inhomogeneous nanoscale system shows that the equilibrium properties (adsorption isotherms, isosteric heats, phase change conditions, etc.) are determined by a small number of dimensionless parameters, including the dimensionless width, *H*, of the nanophase and a *microscopic wetting parameter*, α_w, that is defined in terms of intermolecular forces and solid structure; α_w is a measure of wetting that applies at all scales and for any kind of adsorbed film (gas, liquid or solid) [1,2].

We illustrate the usefulness of this approach using examples drawn from both experimental and molecular simulation studies of gas-liquid and liquid-solid phase separations [3,4], and pressure enhancement in nanopores [5-7], with emphasis on simple pore geometries. These examples illustrate the central role played by wetting, and also the breakdown of some concepts and macroscopic laws, such as the equations of Kelvin, Laplace and Gibbs-Thomson, for nano-phases confined within small pores. They also suggest consistency tests for experimental data that should be applied in the event of capillary condensation or other phase changes in mesoporous materials.

The approach can be extended to account for geometric or energetic wall roughness [8], and to adsorbate molecules that interact with site-site potentials having both Lennard-Jones and point charge sites.

1. R .Radhakrishnan, K.E. Gubbins and M. Śliwinska-Bartkowiak, "Global Phase Diagrams for Freezing in Porous Media", *Journal of Chemical Physics*, **116**, 1147-1155 (2002).
2. Keith E. Gubbins, Yun Long and Malgorzata Sliwinska-Bartkowiak, "Thermodynamics of Confined Nano-Phases", *Journal of Chemical Thermodynamics*, **74**, 169-183 (2014).
3. L.D. Gelb, K.E. Gubbins, R. Radhakrishnan and M. Sliwinska-Bartkowiak, "Phase Separation in Confined Systems", *Reports on Progress in Physics*, **62**, 1573-1659 (1999).
4. Ç. Alba-Simionesco, B. Coasne, G. Dosseh, G. Dudziak, K.E. Gubbins, R. Radhakrishnan and M. Śliwinska-Bartkowiak, "Effects of Confinement on Freezing and Melting", *Journal of Physics: Condensed Matter*, **18**, R15-R68 (2006).
5. Yun Long, Jeremy C. Palmer, Benoit Coasne, Małgorzata Śliwinska-Bartkowiak and Keith E. Gubbins, "Pressure enhancement in carbon nanopores: A major confinement effect", *Physical Chemistry Chemical Physics*, **13**, 17163-17170 (2011).
6. H. Drozdowski, M. Kempinski, M. Śliwinska-Bartkowiak, M. Jazdzewska, Y. Long, J.C. Palmer and K.E. Gubbins, „Structural Analysis of the Behavior of Water Adsorbed in Activated Carbon Fibers", *Physical Chemistry Chemical Physics*, **14**, 7145-7153 (2012).
7. Yun Long, Jeremy C. Palmer, Benoit Coasne, Małgorzata Śliwinska-Bartkowiak, George Jackson, Erich A. Müller and Keith E. Gubbins, "On the Molecular Origin of High Pressure Effects in Nanoconfinement: Effects of Surface Chemistry and Roughness", *Journal of Chemical Physics*, **139**, 144701 (2013).
8. M. Śliwinska-Bartkowiak, A. Sterczyńska, Y. Long and K.E. Gubbins, "Influence of Microroughness on the Wetting Properties of Nano-Porous Silica Matrices", *Molecular Physics*, **112**, 2365-2371 (2014).

464668 Heat Induced and UV Induced Grafting of Poly(glycidyl methacrylate) on PBT Nonwovens for Bioseparations

Abstract:

Polybutylene terephthalate (PBT) nonwovens were successfully grafted with poly(GMA) using a heat induced grafting (HIG) approach using the thermal initiator benzoyl peroxide. The grafting method resulted in uniform and conformal grafted layers around the PBT fibers that were functionalized as anion and cation exchangers for protein capture. Equilibrium protein binding capacities as high as 200 mg of protein per gram of fabric were observed. The equilibrium binding capacities for protein adsorption to the HIG grafted materials were reached within minutes, compared to UV grafted polyGMA ion exchange fabrics which reached equilibrium protein binding capacities in hours, even though they had the same weight % of grafted layers. However, UV grafted ion exchange nonwoven fabrics were capable of binding between 5 to 7 times more protein per mass of fabric at equilibrium than the HIG materials. It was found that the HIG materials showed a decrease in binding capacity with increased molecular weight of the solute that was much steeper than the drop observed with UV grafted materials. These observations indicate that even though the weight % of grafting is the same for HIG and UV grafting, the HIG grafted layers are significantly thinner and have a lower porosity in solution than the UV grafted brush layers.

466177 Thermodynamics and Statistical Mechanics of Drying at the Nano-Scale

Abstract:

Hydrophobically-induced drying transitions at the nano-scale are believed to play an important role in protein folding and transport in ion channels. In this talk I will present recent results on the thermodynamics and statistical mechanics of drying transitions in the presence of flexible hydrophobic substrates. The results suggest that substrate flexibility can play a significant role in both the thermodynamics and kinetics of this process, increasing both the driving force and the rate with respect to the corresponding rigid substrate reference case. The calculations to be discussed include analytical derivations for the underlying thermodynamics, as well as path-sampling simulations for rate calculations and free energy barrier determination. The path-sampling simulations also offer useful insights into the evolution of the evaporation mechanism as a function of substrate flexibility.

471679 Harmonically-Mapped Averaging Applied to Lennard-Jones Crystal Phase

Abstract:

Mapped averaging is a technique we developed to improve ensemble averages during a molecular simulation[1]. When applied to crystals, we map based on the behavior of a harmonic description of the solid and compute the anharmonic contribution to properties of interest with high precision. We can then add in the harmonic contribution and obtain the actual property value. The anharmonic contribution is insensitive to system size and truncation effects while the harmonic contribution can be computed for very large systems with long truncation. The resulting property values are both very precise and accurate with relatively small computational effort.

We have applied mapped averaging to collect potential energy and pressure data for the Lennard-Jones model in the FCC and HCP phases. We fit an equation of state to each set of data to obtain a description of both phases from low to high density (the soft-sphere limit) and up to melting temperatures. We find the HCP phase to be more stable at low temperatures and densities. Coupled with the virial equation of state for the vapor and an equation of state fit to more limited simulation data for the liquid phase, we predict the phase boundaries including the triple point with unprecedented accuracy and precision. We also discuss the effect of vacancies on the properties and phase diagram.

(1) Moustafa, S. G.; Schultz, A. J.; and Kofke, D. A. Very fast averaging of thermal properties of crystals by molecular simulation, Phys. Rev. E 2015, 92, 043303.

458803 Computational-Hydrodynamic Studies of the Noh Compressible Flow Problem Using Non-Ideal Equations of State

Abstract:

The â€œNoh problemâ€ is classic verification problem in the field of computational hydrodynamics of compressible flows. A strong outward facing shockwave is created by impinging a uniformly flowing ideal gas on a hard wall (or imploding it onto the axis of a cylinder or the center of a sphere, depending on the geometry of interest). A simple problem to conceptualize and one which admits an exact analytical solution, it is nonetheless devilishly difficult for numerical codes to predict correctly, making it an ideal code-verification test bed. In its original incarnation, and in nearly all applications since, the fluid is a simple ideal gas; once validated, however, these codes are often used to study highly non-ideal fluids and solids. In this work the classic Noh problem is extended beyond the commonly-studied polytropic ideal gas to more realistic equations of state (EOS) including the stiff gas, the Nobel-Abel gas, and the Carnahan-Starling hard-sphere fluid, thus enabling verification studies to be performed on more physically-realistic fluids. Self-similarity methods are used to solve the Euler compressible flow equations, which in combination with the Rankine-Hugoniot jump conditions, provide a tractable general solution. In the planar case, this solution can be applied to any equation of state and does not necessarily have to exhibit strong shocks; for cylindrical and spherical geometries it is necessary that the analysis be restricted to strong shocks. The exact solutions are compared with numerical results obtained from the Lagrangian hydrocode FLAG, developed at Los Alamos. For these more realistic EOSs, the simulation errors decreased in magnitude both at the origin and at the shock, but also spread more broadly about these points compared to the ideal EOS. The overall spacial convergence rate remained first order. (LA-UR-16-23103)

465378 Lattice Models, Polymers and DNA Topology

Abstract:

The standard Statistical Mechanics models for linear and ring polymers in dilute solution are respectively the self-avoiding walk (SAW) and self-avoiding polygon (SAP) lattice models. For a SAP model, a ring polymer configuration is represented by a random polygon on a lattice such as the simple cubic lattice, and under good solvent conditions, each equal-length polygon is considered to be an equally likely configuration. These models have the advantages that the excluded volume property is easily incorporated and that combinatorial and asymptotic analysis is possible; they have proved to be very useful for understanding critical phenomena, such as phase transitions, for polymer systems. At the same time, despite the simplicity of the models, there are many challenging questions that remain open about them. In this talk, I will review progress made regarding using SAP models to address questions about the entanglement complexity (knotting and linking) of polymer systems. The work has been motivated in part by experimental studies of: (i) enzyme action on DNA, where observed knot and link frequencies characterize the action of the enzyme, and (ii) DNA confined to viral capsids, where observed knot frequencies characterize how the DNA is packed in the capsid.

464879 Design of Macromolecular Biomaterials Using Atomistic and Coarse-Grained Molecular Simulations

Abstract:

In this talk, I will present our recent work using molecular simulations to design novel oligonucleic acids (ONAs) with varying backbone chemistries to tune ONA hybridization and melting. Since double stranded DNA (ds-DNA) is the basis of various bio- and nano- technologies, the need for cheaper alternatives and significant synthetic advances have led to the design of DNA mimics with new backbone chemistries (e.g. click nucleic acids, locked nucleic acids). A fundamental understanding of how these backbone modifications in the oligo-nucleic acids impact the hybridization and melting behavior of the double strands is still lacking. We use coarse-grained (CG) and atomistic molecular dynamics simulations along with well-tempered metadynamics calculations to capture the effects of varying backbone chemistries on the H-bonding between complementary nucleobases and the intra-strand base-base stacking, and ONA hybridization/melting. I will also present our recent results on how conjugation of these novel ONAs with biocompatible polymers (e.g. PEG) impacts the polymer-conjugated ONA hybridization. These polymer-conjugated ONAs will serve as building blocks for thermo-responsive gels and networks in biomaterials.

468481 Rational Design of Peptide-Based Stimuli-Responsive Biomaterials Via Multiscale Modeling

Abstract:

Dynamic, self-assembling peptide-based nanomaterials that can change shape and structure in response to environmental stimuli hold promise to revolutionize medicine and biotechnology. However, the current discovery process is slow and often serendipitous due to the enormously large design space and lack of systematic knowledge as well as predictive models. To tackle the challenge and harness the full potential of â€œsmartâ€ materials, we have built a platform of computer-aided design of nanomaterials by performing *in silico* discovery in collaboration with *in vitro* and *in vivo* experimentalists in facilitating rapid development of novel stimuli-responsive nanomaterials for biomedical applications. Towards this goal, we integrate multi-scale modeling techniques to design and discover novel peptide-polymer conjugates called peptide amphiphiles (PA, comprised of a bioactive peptide sequence and an alkyl tail) that can self-assemble into nanostructures upon a change in a physiological stimulus. We have developed a fundamental understanding of how environmental conditions impact the formation, functionality and performance of these biomaterials.

475310 Introduction to Altsep: Sustainable Separation Processes Roadmap Review

Abstract:

Creating a sustainable future requires sustainable manufacturing processes. This is especially true for the separation processes central to chemical manufacturing. Conventional fluid separations account for over a third of energy use and over half of the capital investment of chemical plants. To pave the way toward transforming the energy efficiency of fluid separations, the ACS Green Chemistry Institute (GCI) has partnered with AIChE to coordinate the development of a NIST-sponsored innovation roadmap. Based on a series of workshops, the draft AltSep roadmap has identified a number of research, development, and demonstration initiatives. These initiatives are critical for establishing a molecular property-based system for the industrial selection, simulation, and design of energy-efficient separation processes and accelerating the industrial application of such alternative separation processes.

A series of speakers will present on the following topics:

- Background and need for the roadmap,
- System to synthesize and design alternative separation processes, and
- Challenges to overcome barriers to adoption of sustainable separation solutions.

Following these presentations, feedback will be solicited from the audience to improve the quality of the roadmap findings.

465142 Slip Slidin' Away: Three Decades of Adventures in Computational Rheology and Lubrication

Abstract:

Carol Hall has been a defining influence on the lives and careers of many of us. As an academic younger brother (Carol received her PhD under the direction of George Stell, with whom Cummings would later perform postdoctoral research), Cummings has enjoyed career-long benefit of Carol's advice and unwavering support. As a young assistant professor new to the discipline that Carol helped him to enter, Cummings sought Carol's advice about where to make a unique impact in chemical engineering. Transport properties was one of the areas that came to the forefront of these discussions, and so began – with the first paper published in 1986 – a three decades long focus on applying non-equilibrium molecular simulation to the prediction and understanding of transport properties in systems ranging from the simple to the complex. We will briefly review some of the highlights of this work, concluding with our recent work exploring the hypothesized hydration lubrication mechanism in knee joints that purports to explain extremely low friction coefficients in a biosystem subject to very high normal pressures.

471438 How to Make Particles Interact in Multiple Directions and Form Interconnected Networks: Colloidal Assembly Based on Field-Induced Multipoles

Abstract:

The assembly of new materials based on particle networks and ordered phases requires the application of "interaction toolbox" that allows independent control of the particle interactions in multiple directions by external field(s). It also necessitates a closely coordinated experimental and theoretical effort, as the enormous parameter space of particle types, interactions, densities, equilibrium/non-equilibrium assembled states and corresponding dynamics are nearly impossible to explore by experimental means only. We will present the experimental and theoretical basis of two strategies for such multidirectional assembly. The first strategy is based on using complex asymmetric particles. Metallo-dielectric particles, such as Janus and patchy spheres and cubes, acquire complex polarization patterns in AC electric and magnetic fields, which can lead to multidirectional interactions. They can be controlled through the field frequency and strength. We will demonstrate how quadrupolar, hexapolar and multipolar polarization interactions induced by external electric and magnetic fields can be used to assemble new types of staggered chains, networks of unusual symmetry and crystals. We will show how the particle assembly motifs and long-scale structures can be analyzed via numerical modeling. The second strategy for multidirectional assembly is to use simple particles that polarize and interact in complex external fields. We show how this can be achieved by simultaneous application of electric and magnetic fields to dispersions of superparamagentic microspheres. We investigate the dynamics of the assembly process, and identify and classify the intermediate states formed. The concurrent field application lead to the formation of bidirectional chains, colloidal networks, and discrete crystals. The morphologies of these structures are in good correlation with theoretical predictions based on Brownian Dynamics simulations. We distinguish two separate kinetic routes for bidirectional assembly, colloidal "polymerization" and "re-polymerization", leading to the formation of structures with specific symmetry, connectivity, directional response or unusual transport properties.

474934 Polymers in Ionic Liquids

Abstract:

Ionic liquids have generated considerable excitement for their varied potential applications and their interesting physical properties. The viability of ionic liquids (ILs) in materials applications is limited by their lack of mechanical integrity, which may be provided by mixing them with a polymeric material. Recent experiments on polymers in ILs have unearthed a wealth of interesting phenomena that raise fundamental questions. This talk focuses on computational studies of PEO in imidazolium ILs. We develop a physically motivated first principles force field; this force field is in quantitative agreement with experiment with no adjustable parameters. The calculations emphasize the importance of polarization but show little evidence for charge transfer. Based on the same quantum calculations we develop a hierarchy of united atom models with decreasing resolution and increasing computational efficiency. Atomistic simulations show a lower critical solution temperature in the PEO in [BMIM] [BF_4] system and provide physical insight into the mechanism based on entropic effects.

454964 *Collective Dynamics of Dipolar and Multipolar Colloids: From Passive to Active Systems*

Abstract:

In this talk I will present recent research on the collective dynamical behavior of colloids with dipolar or multipolar interactions. Starting from "passive" dipolar particles exhibiting network formation and various types of cluster aggregation, we will then move towards systems composed of dipolar "microswimmers" with unidirectional propulsion. These systems display a transition towards a laned state reminiscent of lane formation in driven systems with isotropic repulsive interactions. We interpret our findings by a phase separation process and force balance arguments. Finally, we discuss first results on the collective behavior of dipolar particles with random self-propulsion.

462510 Promising Vistas for Applied Chemical Thermodynamics: Statistical Mechanics As a First Step Toward Thermodynamic Properties

Abstract:

When statistical mechanics is coupled with concepts from physical chemistry, useful results for condensed systems can often be obtained; perhaps the best-known example is Floryâ€˜s derivation of a remarkably sample formula for the configurational entropy of mixing a long-chain polymer with a solvent. For obtaining quantitative thermodynamic properties, statistical mechanics alone is not sufficient; it is also essential to construct a carefully-considered physical model that is simple enough for easy calculations but also sufficiently realistic to represent physical phenomena.

While this procedure is not new, during the last generation, its application for chemical engineering has grown extensively in part, due to the leadership of Carol Hall. To illustrate, some examples from the literature are reviewed. While these examples may or may not refer directly to Carolâ€™s published work, they all exhibit the beneficial influence of her research philosophy.

472566 A Biased Review about Computer Simulations of Protein Folding and Aggregation

Abstract:

Amyloid fibrils are ordered aggregates of misfolded peptides or proteins associated with type 2 diabetes mellitus, Alzheimer's disease, Parkinson's disease, etc. Amyloid diseases are apparently unconnected; however, intermolecular secondary structure (mostly β-sheets) is present in all amyloid aggregates.

Assumptions about the dominance of a particular type of force are common in computer simulations of protein aggregation –most of the times these simplifications are based on experimental observations. However, sometimes the simplifications are rooted on the knowledge of other systems or simply on accepted rules of thumb. Two common simplifications are to overemphasize electrostatic interactions (at one end) or hydrophobic interactions (at the other end). Even though the promoters of each approach make good points in defending them, they (for the most part) ignore the role that solvent may have in stabilizing the structure. Most simulations (are performed with very few peptides, a single protein, a single monomer added to an existing fibril. The fact that aggregation happens at multiple time scales adds complexity to the theoretical work. Prof. Hall's group has used a simplified model for peptides consisting of a few united atoms bearing a square-well potential that represents either hydrogen bonding or hydrophobic forces to study the self-assembly of peptides of alanine and glutamic acid. It is fascinating that such a simple model predicts the correct assembly, even after incorporating the solvent into an oversimplified potential of mean force. Another mechanism that has been explored is molecular crowding; this approach allows for the use of a peptide at infinite dilution in a crowding media. These studies support the idea of incorporating interfaces --we have already shown that interfaces act as an enrichment site in the study of peptide aggregation.

The few reports devoted to study the effect of surfaces on amyloid deposits formation are mostly dedicated to finding a link between the interaction of the deposits with cell surfaces and their toxicity. Concentration

and misfolding of the peptides at the solid/liquid interface support the hypothesis that the surface serves as a catalyst. Recently, we have studied the fibrillation of a series of diblock peptides in the presence of poly(styrene) latex with various surface chemistries and of liposomes with different phospholipid compositions. It was concluded that fibrillation depends on the composition of the diblock and on the presence or absence of interfaces but no adsorption was observed. Interestingly, the replacing of an isoleucine block by a leucine one in diblock peptides of the same length inhibits fibrillation. The consequences of these findings on current modeling efforts as well as the development of a new model are discussed.

464055 “Grafting through” Polymerization

Abstract:

We use a combination of computer simulation and experiments to comprehend “grafting through” polymerization, i.e., copolymerization of free monomers in solution and polymerizable units bound chemically to a substrate. Free polymer chains are formed initially in bulk solution and if they incorporate at least one surface-bound monomer, the macromolecules get covalently bonded to the surface during the polymerization process. We concentrate specifically on addressing the effect of spatial density of the surface-bound monomers on the formation of the surface-attached polymers. We employ a lattice-based Monte Carlo model based on the bond fluctuation model scheme to provide molecular-level insight into the grafting process. For experimental validation, we create gradients of density of bound methacrylate units on flat silicon wafers using organosilane chemistry and carry out “grafting through” free radical polymerization initiated in bulk. The proximity of the surface-bound polymerizable units promotes the “grafting through” process but prevents more free growing chains to “graft through" the polymerizable units.

468105 Understanding Temperature-Responsive Behavior of Elastin-like Polypeptides

Abstract:

The elastin-like polypeptide (ELP) is temperature sensitive biopolymer which undergoes conformational transition at critical inverse transition temperature or LCST; the feature which has led to their use in a variety of applications. ELPs are defined by a pseudo-periodic, low complexity sequence with repeat motifs, as in a model system with repeat sequence of VPGXG. Substitution of X residue with various amino acids or modifications of chain length and solute quality allows for tuning the inverse transition temperature. However, the molecular origin of LCST and the observed differences as a function of various factors are not yet completely understood. In order to elucidate the factors driving the conformational transitions of ELPs we performed large scale all-atom molecular dynamics simulations of ELPs with different sequence and length and at a range of temperatures. Comprehensive analyses of the hydrophobicity, shape, size, hydration and dynamics of ELPs suggest that the transition behavior includes the thermal disruption of water network, loss of hydration, and increase in peptide hydrophobicity. In this talk I will discuss the temperature sensitive behavior of polypeptides as a function of sequence, sequence direction, length, and concentration. The results from our study provide an atomic-level description of the thermo-responsive conformational properties of elastin-like peptides and explain experimental observations.

477518 Utilizing Multivariate Analysis to Enhance Bioprocess Development and Operation

Abstract:

Advances in the development of cell culture bioprocesses have resulted in an extraordinary amount of data, which spreads over all stages of development and operation. This accumulation of data is commonly overlooked but often contains untapped resources and information. This talk aims to outline high level strategic approaches for multivariate analysis and visualization techniques which can extract this knowledge and be further applied to enhance all facets of biopharmaceutical production.

477948 Self-Assembly of Colloids in Ordered Carrier Fluids

Abstract:

The talk focuses on the complex interplay of molecular scale effects, nonlinearity of the orientational field, and long-range elastic forces which make the liquid-crystal physics very challenging. A consistent way to extract information from the microscopic, molecular scale up to the meso- and macroscopic scale is still missing. Here, we develop a hybrid procedure that bridges this gap by combining extensive Monte Carlo simulations, a local Landau-de Gennes theory, classical density functional theory, and finite-size scaling theory. As a test case to demonstrate the power and validity of our novel approach we study the effective interaction among colloids with Boojum defect topology immersed in a nematic liquid crystal. In particular, at sufficiently small separations colloids attract each other if the angle between their center-of-mass distance vector and the far-field nematic director is about 30ï,°. Using the effective potential in coarse-grained two-dimensional MC simulations we show that self-assembled structures formed by the colloids are in excellent agreement with experimental data.

450819 Role of Van Der Waals and Entropy on Gold Cluster (meta)Stability

Abstract:

Discerning the ground state structure of gas phase gold clusters, and in particular the critical size when gold clusters begin favoring three-dimensional (3D) structures over two-dimensional (2D) structures at 0 K, has been a topic of longstanding interest. At finite temperature, however, gold clusters exhibit many structural isomers of similar energies and can interconvert between them. To move beyond the monostructure description of clusters at 0 K, we performed *ab initio* replica-exchange molecular dynamics (REMD) on Au_n clusters (n =5-14) to identify metastable states, their relative populations between 100-400 K, and to examine the influence of entropy and van der Waals (vdW) on the isomer energetic ordering. REMD calculations used PBE with the scalar ZORA relativistic correction and Many-Body-Dispersion (MBD) vdW. Free energies of Au_n isomers are optimally estimated using the Multistate Bennett Acceptance Ratio, and the 2D and 3D isomer populations are examined at different temperatures and cluster sizes. The distribution of bond coordination numbers and radius of gyration are used to address the challenge of discriminating isomers along the dynamical trajectories. Dispersion effects are shown to be critical to obtain the correct isomer energetic ordering, stabilizing 3D structures relative to 2D structures by up to 0.4 eV. Entropic effects typically stabilize metastable 3D structures relative to planar/quasiplanar structures. Computed IR spectra of Au_n are compared with experimental spectra obtained via far-IR multiple photon dissociation in a molecular beam at 100 K, which enables the structures of Au_9, Au_{10}, and Au_{12} to be assigned.

453957 Simulating NMR Relaxation Time Constants of Benzene in MOF-5 to Investigate Phase Coexistence

Abstract:

Capillary condensation has long been thought to be impossible in a microporous framework due to the small pore size making the system analogous to a one-dimensional Ising model. Steps in the adsorption isotherms of MOF-5, a model microporous metal-organic framework, have thus been difficult to explain as they are reminiscent of capillary condensation. Recently, molecular simulations have shown bulk-like vapor and liquid phases of benzene coexisting in MOF-5 with the domains extending across multiple unit cells, explaining these puzzling results. ^{1}H NMR Relaxometry has provided experimental evidence for this phase coexistence as multiple relaxation time constants have been found for a sample of benzene in MOF-5, which have been hypothesized to be due to the multiple phases. We have conducted molecular dynamics simulations of the benzene in MOF-5 system and calculated the resulting NMR relaxation times; the time constants have been separated into contributions due to dipole-dipole interactions that are intramolecular between protons on a benzene molecule, intermolecular between the protons on different benzene molecules, and intermolecular between the benzene protons and the MOF-5 protons. These contributions can also be experimentally determined by using various amounts of deuterated benzene and deuterated MOF-5 to limit the effects of the various interactions.

454654 Supramolecular Polymer Hydrogel Design through Molecular Dynamics Study of Octa-Acid Cavitand and Adamantane

Abstract:

Supramolecular polymer (SMP) hydrogels are an exciting class of smart materials, which regenerate through guest/host connections incorporated into the polymer network. [1] Such materials are biodegradable, lightweight, resilient and durable and could redefine our ability to conserve global resources. An important step to discovering a polymer hydrogel thermodynamically driven to regenerate is study of the hydrophobic effect – the direction of guest and host as determined by aqueous solvent – and its ability to direct guest/host pairings. Many classes of organic chemicals such as cyclodextrins, deep octa-acid cavitands (OA) and curcubiturils have been shown to have a strongly hydrophobic interior with a hydrophobic exterior, making them suitable for hydrophobic directed assembly with any hydrocarbon. [2],[3] Existing polymer simulations focus on adamantine as a potential guest, and our experimentation focuses on adamantine and OA derivatives.[4] Furthermore, our MD experiments between OA and adamantine have shown co-nonsolvency effects, where water and ethanol mixtures of ~2-5% have been shown to create a local binding maximum. This novel finding suggests competition between the solvent molecules for sites in the OA interior and could explain the origins of "high energy water" in which incomplete hydrogen binding networks alter the local thermodynamics of water inside the OA cavity. [5]

---

[1] Akira, H, et al. "Macroscopic self-assembly through molecular regocnition." Nature Chemistry. **2011,** *3*(1), 34-37.

[2] Takahashi, K. "Organic reactions mediated by cyclodextrins." Chem. Rev. **1998**, 98, 2013–2033

[3] Wanjari, P.P, B.C. Gibb, and H.S. Ashbaugh. "Simulation optimization of spherical non-polar guest recognition by deep-cavity cavitands." The Journal of Chemical Physics, **2013,** *139*, 234502

[4] Wang, S., C.C. Chen, and E.E. Dormidontova. "Reversible association and network formation in 3:1 ligand-metal polymer solutions." Soft Matter, **2008**, *4*. 2039-2053

[5] Biederman, F, et al. "The Hydrophobic Effect revisited – studies with supramolecular complexes imply high-energy water as a non covalent driving force." Angew. Chem. Int. Ed. **2014.** 53, 11158-11171

453660 Resolving Experimentally-Inferred Solvophobic Interactions in the Overlap Region

Abstract:

Solvent-mediated interactions between non-polar atoms is a principal driving force in molecular self-assembly. Experimental measurement of two atoms' solvent-mediated interactions is difficult; such a measurement when two atoms are overlapping is impossible. In the 1970's, however, Ben-Naim derived useful experimental estimates of the solvent-mediated contribution of the potential-of-mean-force (PMF) for two methanes at distances within the overlap region. By approximating two methanes separated by 1.54 Å—the bond length of an ethane—as a single ethane, he could calculate the free energy of solvation of both ethane and methane, and therefore calculate the potential of mean force at 1.54 Å as: $\mu^{ex}{}_{ethane} - 2\mu^{ex}{}_{methane}$.

In this study we use simulations to revisit Ben-Naim's results. By using previous work on interpolating and calculating the solvent-mediated interactions of methanes in the overlap region, we are able to quantitatively compare simulation results with experimental estimates. We compare methane interaction free energies, entropies, and enthalpies—in water and in ethanol—with Ben Naim's estimates. By using accurate force field models—TraPPE-UA and TIP4P2005—we are able to reproduce Ben-Naim's experimental estimates. We observe, however, that such results understimate the solvent-mediated contribution to the PMF in the overlap region. When we perform a free-energy perturbation on the ethane, transforming the methyl groups into methanes the results become consistent with previous simulation work on solvent-mediated contributions in the overlap region.

456442 Computation of Chemical Potential of Water at Low Temperatures

Abstract:

Computation of Chemical Potential of Water at Low Temperatures

Calculations of free energy changes and chemical potentials are of great importance in chemical engineering. In this work, the performance of the Gibbs ensemble combined with Continuous Fractional Component method (GE-CFCMC) for free energy calculation in dense systems has been compared to the Widom test particle insertion in the conventional Gibbs ensemble. For this study, 3 different water models have been considered: TiP3P-EW, TiP4P-EW, SPC. It is shown that while Widom test particle method fails to provide a correct estimation of the chemical potential for the liquid phase at low temperatures, the simulations using GE-CFCMC can estimate chemical potential correctly. It is also shown that the acceptance probability for the particle exchange between the boxes has been improved using the GE-CFCMC method by Poursaeidesfahani et al. [Direct Free Energy Calculation in the Continuous Fractional Component Gibbs Ensemble, Journal of Chemical Theory and Computation, 2016, 12 (4), pp 1481â€"1490]. The GE-CFCMC method has eliminated one of the coupled fractional particles in one of the boxes and therefore, operates only with a single fractional particle in one of the boxes at a time. Therefore, especially when the fractional particle is in the gas phase, the pair potential calculations in the liquid phase are no longer needed. This reduces the computational time significantly.

456358 Multi-Scale Coarse Graining of Polarizable Models through Force-Matched Dipole Fluctuations

Abstract:

An extension of the multiscale coarse-graining (MS-CG) method to polarizable coarse-grained (CG) models is presented. In the extension, force matching is used to derive charged dimers that mimic the dipole behavior, including electronic polarizability, of fine resolution systems. The extended MS-CG method separates short-range and electrostatic forces, and treats the polarization interactions by representing dipole fluctuations with a single harmonic bond reminiscent of the Drude oscillator. The new method is first tested on several flexible alcohols, where the transferable head groups in the atomistic field lead to transferable electrostatics in the CG field. The method is then applied to a polarizable methanol model, where the CG dimer is able to match the atomistic dipole distribution. All force fields are benchmarked with radial distribution functions, and dielectric constants. The static dielectric constants agree well with atomistic models, but the faster dynamics in the CG ensembles leads to significant deviations in the frequency dependent permittivity. Model transferability is checked by measuring response to static external fields through a dipole coupling order parameter. Effective polarizabilities are inferred from the dimer harmonic bonds, and an *ad hoc* correction is used to improve one CG force fieldâ€™s dipole magnitude response to electric fields.

457734 Desolvation Barriers in Ionic Crystal Growth

Abstract:

Desolvation Barriers in Ionic Crystal Growth

Mark N. Joswiak,[a] Michael F. Doherty,[a] Baron Peters[a,b]

[a] Department of Chemical Engineering, University of California-Santa Barbara

[b] Department of Chemistry and Biochemistry, University of California-Santa Barbara

Crystal habits are dependent on characteristics of the solute, but the solvent can dramatically impact the morphology [1]. The role of the solvent is two-fold â€" altering surface energies and/or attachment barriers. While numerous studies have explored the â€œsolvent effectâ€ in terms of surface energies [2], few studies have explored desolvation/attachment barriers [3-6], which have been implicated as the rate-limiting step for crystal growth. Understanding these kinetic barriers is critical for accurate models of crystal growth and dissolution. We use rare-event simulation techniques and mechanistic hypothesis testing [7] to investigate the role of water molecules on the detachment of ions from kink sites of a NaCl crystal. We are able to characterize the detachment/attachment process in terms of general, simple reaction coordinates involving 1 solute and 1 solvent collective variable. We discuss the similarities and differences between the two ions during the detachment process. From our computed detachment rates, we calculate the rate of NaCl dissolution and find excellent agreement with an independent study [8]. We discuss the applicability of our results to other important and interesting systems, such as calcium carbonate crystal growth.

References:

[1] P. Dandekar, Z.B. Kuvadia, M.F. Doherty, *Annu. Rev. Mater. Res.* **2013**, *43*, 359-386.

[2] S.H. Kim, P. Dandekar, M.A. Lovette, M.F. Doherty, *Cryst. Growth Des.* **2014**, *14*, 2460-2467.

[3] P.G. Vekilov, *Cryst. Growth Des.* **2007**, *7*, 2796-2810.

[4] A.G. Stack, P. Raiteri, J.D. Gale, *J. Am. Chem. Soc.* **2012**, *134*, 11-14.

[5] S. Piana, F. Jones, J.D. Gale, *J. Am. Chem. Soc.* **2006**, *128*, 13568-13574.

[6] L.M. Liu, A. Laio, A. Michaelides, *Phys. Chem. Chem. Phys.* **2011**, *13*, 13162-13166.

[7] B. Peters, *Mol. Simulat.* **2010**, *36*, 1265-1281.

[8] G. Lanaro, G.N. Patey, *J. Phys. Chem. B* **2015**, *119*, 4275-4283.

458367 Mechanism of the Fcc-Hcp Phase Transformation in Solid Ar

Abstract:

We present an atomistic description of the mechanism of the fcc–to–hcp transformation in solid argon obtained from molecular dynamics transition path sampling simulations. We show that the transition barrier at 40 K under ambient pressure is close to the thermal energy per atom. The analysis of the ensemble of reactive trajectories collected during the sampling revealed three transition types. The average coordination number is used as order parameter and to monitor the progress of the transformation. In each transition type, under– and over–coordinated atoms form and distribute uniformly throughout the lattice during the transformation. Stacking disorder is discussed to de- scribe the transition process and the cooperative atomic motions. The high activation energy for collective movements of many atoms explains previous experimental observations of a sluggish bulk transformations and difficulties of observation of the bulk transition at lower temperature and ambient pressure. The transition mechanism in bulk Ar is different from Ar clusters as the orthorhombic intermediate structure proposed for the latter is not observed.

459272 Prediction of the Sooting Tendencies of Candidate Biofuels from Molecular Structure Via an Artificial Neural Network

Abstract:

The tendency of candidate bio-based fuel blendstocks to produce soot in traditional engine designs is a significant constraint in screening compounds for next generation biofuels. Experimental methods have thus been developed for rapidly screening the sooting tendencies of small quantities of compounds. While such methods go a long way in easing the experimental burden associated with testing soot formation, it is difficult to obtain an estimate of a molecules tendency to form soot for expensive or otherwise difficult to obtain molecules. We have therefore analyzed the Yield Sooting Index (YSI) measurements from 295 fuel-like compounds recently compiled by McEnally & Pfefferle (2010) to determine if sooting index can be predicted accurately from molecular structure. Molecular descriptors were calculated for each compound and used to train a feed-forward artificial neural network (FF-ANN). Prior to training, 20% of the dataset was reserved as a test set to validate the predictive capacity of the model. The trained model showed median absolute errors of a similar range to the expected experimental error range, indicating that YSI can indeed be predicted by molecular structure. Furthermore, validations of the network architecture indicate that the model generalizes well to new classes of molecules, and subsequent experimental testing has indicated that the model indeed correctly predicts general sooting trends. This work should provide an easy framework for screening the viability of large lists of potential bio-blendstocks, as high-sooting species can be identified directly from molecular structure.

459170 Radical Mechanism for Low Temperature C-O Bond Activation on Metal/Metal Oxide Catalysts

Abstract:

Selective C-O bond scission is a key step for a variety of industrial processes ranging from Fischer-Tropsch synthesis to valorization of biomass. Recently discovered multifunctional metal/metal oxide catalysts (e.g., Rh/MoO_x, Ir/VO_x) exhibit unusual activity and selectivity for C-O bond hydrogenolysis at temperatures < 200°C. Here, we combine density functional theory calculations, electronic structure analysis and microkinetic modeling to develop comprehensive insights.

We describe a highly efficient mechanism of carbon-oxygen bond activation on metal/metal oxide multifunctional catalysts, which potentially reconciles the contradiction between low C-O bond scission activities, exposed by models and surface science experiments, and high experimentally observed reaction rates. We employ catalytic transfer hydrogenolysis chemistry of furfural to 2-methyl furan on a Ru/RuO_2 catalyst as an illustrative example due to its importance in the production of renewable fuels and chemicals. Unexpectedly, we discover an oxygen vacancy-mediated reverse Mars-van Krevelen-type radical mechanism that explains the high catalyst activity. Radical mechanisms are ubiquitous in various catalytic high temperature oxidations of, e.g. methane or propylene; yet, to our knowledge, this is the first study introducing the radical *reduction* mechanism in heterogeneous catalysis at low temperatures in the liquid phase. We describe electronic properties of both the catalyst and the key reaction intermediate that contribute to facile chemistry and overcome energy scalings. Finally, we find the mechanistic picture to be fully consistent with isotopic labeling experiments and kinetic studies. Given the importance of heterogeneous catalysis for tailored bond-breaking in energy production and storage, we believe that our discovered mechanism provides an important step toward development of low temperature processes catalyzed by surfaces.

The key findings are the following:

1. For the first time, we establish a consistent mechanistic picture of a vacancy formation mechanism on RuO_2 from first principles microkinetic modeling that explains all reactivity trends observed in surface science experiments.
2. We show that RuO_2 vacancies are by far the most reactive sites for activation of the C-O bond and the only sites that are consistent with low temperature activity of the catalysts.
3. Unexpectedly, we discover that the C-O bond scission on a vacancy proceeds via a furfuryl radical intermediate, allowing us to classify it as a **reverse Mars-van Krevelen-type radical reduction mechanism**, which has not been reported previously.
4. We elucidate a single reactivity factor that underlines the radical reduction mechanism and its generality to other catalytic systems. Unlike on metals, where the reaction barrier is lowered due to binding of the final products to the surface, here the low barrier is a result of the enhanced stability of the weakly bound radical due to **conjugation**.

459759 Understanding Unfolded and Intrinsically Disordered Proteins Via Molecular Simulations

Abstract:

Intrinsically disordered proteins (IDPs) do not fold to a single three-dimensional structure under physiological conditions; however, still have biologically relevant functions in the cell. A significant fraction, more than 1/3rd, of eukaryotic proteins are fully or partially disordered. While IDPs in general perform critical functions in the cell, some of them are also closely related with serious human diseases, including Alzheimerâ€™s, Parkinsonâ€™s, and type II diabetes. It is quite difficult to characterize the heterogeneous disordered ensemble of IDPs with available experimental techniques. Molecular simulations offer a unique avenue for better understanding of IDPs and unfolded proteins. This poster will cover our recent molecular dynamics simulations revealing kinetic and equilibrium properties unfolded and disordered proteins.Â

460463 Understanding the Shape-Control Mechanism in the Solution-Phase Growth of Ag Nanocrystals

Abstract:

Nanostructures with well-defined shapes can be synthesized using the solution-phase polyol process in the presence of polyvinylpyrrolidone (PVP), which is known to play a key role in directing shape control. However, the mechanism by which PVP actuates shape control remains elusive. We hypothesize that the influence of PVP on the Ag atom deposition kinetics dominates the shape-control mechanism. To test this hypothesis, we use umbrella sampling molecular dynamics simulations to calculate the potential of mean force (PMF) and diffusion coefficient profiles of Ag atom deposition onto Ag(100) and Ag(111) in ethylene glycol solution with PVP. These profiles are used to calculate the mean-first passage times and implement extensive Brownian dynamics simulations, which allows quantitative evaluation of the process kinetics. Our results show that PVP films can regulate the flux of Ag atoms to be greater towards Ag(111) than Ag(100) facets. Under kinetic control, \{100\}-faceted nanocrystals will be formed when the Ag flux is greater towards Ag(111). The predicted kinetic Wulff shapes are in agreement with the analogous experimental system, which confirms our hypothesis.

460750 Extraction of Pairwise Interaction Potentials from Cohesive Energy Inversions--a General Approach with DFT

Abstract:

Cohesive energy curves contain the energetics of atomic interactions in crystalline materials. These can be obtained from density functional theory calculations using in silico lattice expansions. Decomposing these curves into the different interatomic potential energy contributions is valuable to extract pair-potential interactions, which can later be used for molecular mechanics or molecular dynamics simulations. In this work, we present and discuss a generalized method for the extraction of pairwise interaction potentials from the inversion of cohesive energy curves. We use detailed geometrical descriptions of the atomic interactions to construct a list of atomic displacements and degeneracies, which is modified using a Gaussian elimination process to isolate the pairwise interactions.
The proposed method provides a more general framework for cohesive energy inversions that is robust and accurate for systems well-described by pairwise potential interactions. Results show very good reproduction of cohesive energies with the same or better accuracy than current approaches. An important advantage of this method is its application to systems with fixed internal constraints such as local rigid body structures. Specific examples for the extraction of heteroatomic interactions in metal hexaboride systems are illustrated.

460796 Molecular Dynamics of Inorganic and Organic Interfaces, Parametrization Based on Quantum Mechanical Simulation

Abstract:

Interface between inorganic and organic materials is a key factor in nanocomposites. The structural and energetic properties are important points to understand physics of the composites. The molecular dynamics (MD) is powerful tool to simulate and analyze such system, however, the force-field (FF) parameter set appropriate to the interfacial system is not always accessible. One approach to obtain FF parameter set is to determine the parameters so that the potential surfaces calculated in FF and quantum mechanical (QM) simulation coincide. We perform a hierarchical simulation of QM, full atomistic (FA) MD, and coarse grained (CG) MD. The FF parameter sets for the nonbonding interaction in the FA and the CG models are constructed from the QM simulation result. That for bonding interaction in the CG model is constructed from the FA MD simulation result. In this study, we utilize SIESTA for the QM simulation and J-OCTA for the system modeling and the MD simulation. We will provide some examples of the MD simulations with QM based FF parameters.

459932 Kinetic Analysis for Molecular Simulations of Nucleation Processes with Low Barriers

Abstract:

A nucleation-mediated phase change occurs through the spontaneous formation of clusters of the new phase within the old phase. Due to the competing contributions from the interfacial free energy penalty and the bulk driving force for crystallization, there is a kinetic barrier with respect to the size of the cluster formed. Once a large enough cluster is formed to pass over the kinetic barrier, rapid growth of the new phase occurs. The relative timescales associated with nucleation and growth have important implications for the analysis of nucleation data obtained from molecular dynamics (MD) simulation. We have developed a method for the analysis of data from MD studies of nucleation processes with small separation between timescales of nucleation and growth. The method is well-suited for nucleation events with low kinetic barriers, in addition to large systems since the separation between timescales becomes smaller as the size of the system increases. In our method, we use the mean first-passage time of the largest cluster, obtained from several nucleation runs, in order to parameterize a theoretical mean first-passage time (MFPT) curve based on the Becker-Döring model. In order the make the method computationally practical, we developed an accurate approximation for the first-passage time distribution based on its cumulant expansion. Application of the method to MD data yields estimates for important kinetic parameters including the critical nucleus size, free-energy barrier, and critical nucleus size, as well as the nucleation rate and a characteristic growth rate. The method was applied in a case study on crystal nucleation of *n*-eicosane in the presence of flow fields of varying intensity. We find that the method is able to capture the decrease in the free-energy barrier as the intensity of the flow field is increased.

461392 First-Principles Density Functional Theory Modeling Assisted Understanding on the Redox Properties of Nitrogen Doped Organic Molecules

Abstract:

First-Principles Density Functional Theory Modeling Assisted Understanding on the Redox Properties of Nitrogen doped Organic Molecules

Jiwoong Kang[1, 2], Ki Chul Kim[1], and Seung Soon Jang[1]

[1]School of Materials Science and Engineering, Georgia Institute of Technology, 771 Ferst Drive NW, Atlanta, GA 30332-0245, USA

[2]School of Chemical and Biomolecular Engineering, Georgia Institute of Technology, 311 Ferst Drive NW, Atlanta, GA 30332-0100, USA

Lithium-ion battery has attracted as a candidate for an efficient storage of the electrochemical energy towards the applications to the portable devices and electric vehicles owing to its high performance and cyclic stability. Despite of its potential, the slow diffusion of lithium ions due to the intrinsic property of the cathode materials mainly employing transition metal oxides results in the poor power density and thus impedes the practical applications of the lithium-ion battery. In addition, finding cathode materials with high redox potentials is another critical issue to be resolved for developing high-potential lithium-ion batteries. In this study, we introduced a series of organic molecules in which it is believed the diffusion of the lithium ions would be significantly improved. In this study, we investigated redox chemistry of a series of nitrogen doped organic molecules with different geometry of doped atom. First-principles

density functional theory was employed to stress the geometric effect of doped atoms on redox chemistry. This study will provide extended understanding of strategies used to improve properties of lithium ion battery.

461272 Automated Spatial Analysis of NFκB in Multispectral Images of Cerebrovascular Endothelial Cells

Abstract:

Cerebrovascular endothelial cells play a key part in the inflammatory response of the blood-brain barrier in pathological conditions such as Alzheimer's disease. Specifically, the NF-κB signaling pathway plays a central role. Better understanding of the factors in inflammatory disease progression can lead to more effective treatments for such devastating illnesses like Alzheimer's, asthma, arthritis, cancer, diabetes and many more inflammatory diseases. The presented approach analyzes spatial NF-κB distribution contained in multispectral stacked micrograph images of cerebrovascular endothelial cells indexed based on dose of the activating protein and the length of activation. Image analysis identifies the location of nuclear boundaries and quantifies NF-κB in relation to the closest nuclear boundary. This information is used to develop a mathematical model that describes the time and concentration dependence of NF-κB in response to the activating proteins. The proposed method allows for analysis and modeling of previously unexplored spatial behavior of NF-κB.

461766 A Grand Canonical Monte Carlo Simulation-Based Method for the Calculation of Electrical Potential inside Nano-Porous Electrodes

Abstract:

Excess charge at the surface of an electrode, created via an externally applied potential, results in an accumulation of counterions near the surface and a depletion of coions in that region. Therefore, it creates an excess electric potential at the interface. This potential drops as the distance from the surface to the Galvani potential in the solution bulk. This phenomenon is known as the electrical double layer (EDL). The EDL governs natural phenomena like colloidal stability and electrokinetic phenomena; and affects the outcome of electrochemical reactions. Classical Theory describes the EDL via the combination of the Poisson equation for electrical potential and the Boltzmann distribution of charge within an electric field next to a planar charged solid surface. This approach treats ions as non-interacting point charges in the vicinity of a single-charged surface in contact with a bulk solution. However, most current applications where EDL is relevant involve confinement within nanostructures, where the size of the ions is within one or two orders of magnitude of the size of the structure itself.

We will present a method to describe the EDL within charged slit-type nanopores that combines Grand Canonical Monte Carlo simulations (GCMC) with electrodynamics concepts to determine not only the EDL structure, but most importantly the profiles of electrical potential. The method was validated against the simplest case where Classical EDL Theory can be applied (two parallel planar surfaces with no EDL overlap) and then extended to increasingly confined spaces. In contrast to classical EDL theory, this method accounts for ion-size and ion-ion interactions; and provides insights into the EDL properties within nanostructures.

461827 Meta-Analysis of Cellular Toxicity of Cadmium-Containing Quantum Dots Using Bayesian Networks

Abstract:

The rapidly expanding application of engineered nanomaterials (ENMs) across diverse areas, such as medicine, electronics, biomaterials, and energy production, raises concerns about their potential adverse impacts on the environment and health. Accordingly, there is a global drive to ensure that the ENMs developed for various purposes will not pose risk to human health and the environment. In this regards, voluminous data have been generated in high throughput toxicity screening studies to evaluate bacterial and cellular bioactivity induced by exposure to ENMs. Reported toxicity data can be structured or unstructured, quantitative or qualitative, and at different levels of confidence and consistency across different toxicological studies. Currently, efforts to arrive at generalization of ENM toxicity behavior *via* data-driven models have typically been based on datasets from single studies with little focus on the collective body of published evidence. In general, irrespective of the level of complexity of sought toxicity models, the application of literature data mining and knowledge extraction approaches to heterogeneous datasets must consider the value of information and the body of evidence in order to: (1) evaluate the relevance and significance of various ENMs physicochemical and experimental conditions (i.e., attributes) to toxicity metrics, and (2) develop data-driven models for correlating toxicity metrics with the identified quantitative and qualitative attributes. Accordingly, the present study presents a cause-effect inference approach, suitable for analyzing and extracting knowledge from published body of evidence, focusing on the cellular toxicity of cadmium-containing semiconductor quantum dots (QD). In the present approach Bayesian Networks (BNs) were developed using 2,703 data samples (each with 24 qualitative and quantitative attributes) collected from 448 publications. The BN serve as both a data-driven model for predictive to QD cellular toxicity but can also provide information about the conditional dependence of the toxicity on various ENM properties and exposure conditions. Analyses of the body of evidence via BNs demonstrated that cellular toxicity of QDs closely correlated with their surface properties (including shell, surface charge, ligand and surface modification), diameter, assay type and exposure time. The present work suggests that BNs are effective for integrating

qualitative and quantitative data thereby providing the means for interrogating wide-range of toxicity data and for use in predictive toxicology via meta-analysis.

462049 Effects of Self-Energy of the Ions on the Double Layer Structure and Properties at the Dielectric Interface

Abstract:

Although numerous theoretical efforts have been put forth, a systematic, unified and predictive theoretical framework that is able to capture all the essential physics of the interfacial behaviors of ions, such as the Hofmeister series effect, Jones-Ray effect and the salt effect on the bubble coalescence remains an outstanding challenge. The most common approach to treating electrostatic interactions in the presence of salt ions is the Poisson-Boltzmann (PB) theory. However, there are many systems for which the PB theory fails to offer even a qualitative explanation of the behavior, especially for ions distributed in the vicinity of an interface with dielectric contrast between the two medium (like the water-vapor/oil interface). A key factor missing in the PB theory is the self energy of the ion.

We develop a self-consistent theory that treats the electrostatic self energy (including both the short-range Born solvation energy and the long-range image charge interactions), the nonelectrostatic contribution of the self energy, the ion-ion correlation and the screening effect systematically in a single framework. The theory gives a continuous self energy across the interface, which allows ions on the water side and the vapor/oil side of the interface to be treated in a unified framework. The theory involves a minimum set of parameters of the ion, such as the valency, radius, polarizability of the ions, and the dielectric constants of the medium, that are both intrinsic and readily available.

Using the theory, we demonstrate three essential effects, the image charge effect, the inhomogeneous screening effect and the specific ion effect, for the ions near the dielectric interfaces. First, we show that the image charge repulsion creates a depletion boundary layer that cannot be captured by a regular perturbation approach. The image force qualitatively alters the double layer structure and properties, and gives rise to many non-PB effects, such as nonmonotonic dependence of the surface energy on concentration, like-charge attraction and charge inversion. Then, we show that the double layer structure and interfacial properties is drastically affected by the inhomogeneous screening due to the nonuniform ion distribution. At higher concentrations when the bulk Debye screening length is comparable to the Bjerrum length, the double layer structure and interfacial properties are significantly affected by the inhomogeneous screening. In particular, the depletion zone is considerably wider than that predicted by the bulk screening approximation or the WKB approximation. The characteristic length of the depletion layer in this regime scales with the Bjerrum length, resulting in a linear increase of the negative adsorption of ions with concentration, in agreement with experiments. Finally, we study the self energy of a single ion across the dielectric interface. Using intrinsic parameters of the ions, such as the valency, radius, and polarizability, we predict the specific ion effect on the interfacial affinity of halogen anions at the water/air interface, and the strong adsorption of hydrophobic ions at the water/oil interface, in agreement with experiments and atomistic simulations.

461890 Predicting Protein Dynamics in Complex Environments

Abstract:

A simulation model for proteins at the mass transport scale has been developed for the purpose of studying the separation of proteins based on isoelectric point and ionic strength. Coarse graining is required to investigate the mass trasport regimes of proteins using computational methods. The raspberry model for colloid particles is a rigid body defined by generic beads spready evenly over the surface. The hydrodynamics are resolved using lattice Boltzmann fluid. Spherical and ellipsoidal geometries have been constructed to validate the hydrodynamic characteristics of these colloid particles, namely, the effects of size, shape, and confinement on the diffusion tensor. The model has been extended to protein-shaped particles using the solvent accessible surface of a protein and charges assignment based on electrostatic potential mapping of the atomistic protein model. The computation effort is vastly reduced while maintaining accurate descriptions of the excluded volume, electrostatic, and hydrodynamic effects that dominate protein transport.

462448 Automating Workflows for Surface Science and Catalysis

Abstract:

Density functional theory is rapidly becoming integral to the investigation of materials properties at the atomic scale. In chemical engineering, DFT has been particularly successful in the simulation of chemical surfaces for catalysis and adsorption, having aided in explaining various phenomena related to catalytic activity and even predicted new catalysts which have been confirmed in experiment. While DFT-based tools have become significantly more advanced in recent years, most scientific workflows for determining catalytic activity using electronic structure calculations still require significant effort by users to manage individual DFT simulations, parse output, and analyze the resultant data.

In this work, we present recent progress in the development of a high-throughput workflow to yield adsorption energies of low-index facets of a given material. Using the Materials Project infrastructure and approach, we demonstrate that simple analytical tools for the generation of surfaces and placement of adsorbates can enable more efficient automation of DFT calculations. In addition, our workflow automates collection of results from electronic structure simulations and performs a preliminary analysis in order to construct a database of properties that can be compared to experiment and filtered to identify promising candidate materials for further study. Perhaps most importantly, our workflow is designed to process materials and adsorbates in such a way as to minimize the necessity for user intervention, even for complex materials, and thus enable a more simple approach to comprehensive analysis. Lastly, we demonstrate how data collected using our workflow can be leveraged into insights into applications in nanoparticle synthesis and shape prediction.

463992 Potential of Mean Force of siRNA and Pmal: Molecular Dynamics and Theoretical Analysis

Abstract:

Effective delivery of siRNA is a significant problem which involves the health of human beings[1-2]. PMAL (Poly (maleic anhydride-alt-1-decene) substituted with 3-(Dimethylamino) propylamine) is introduced to facilitate siRNA delivery because of amphipathy and biological compatibility[3-4]. The interaction between siRNA and the carrier is the basic problem that determines stability of complex and transmembrane process[5-6]. However, itâ€™s still lack of theory describing the complexation process though many simulation results elaborate the interaction of siRNA and carrier[7]. In this work, the effect of charge of PMAL on PMF curves is investigated with molecular dynamics. Based on the MD results, Manning theory is used to explain counterion condensation while Debye-Huckel approximation is employed to predict PMF. The extent of theory will be broadened by the success of this case and itâ€™s a sign that simple theory can be applied to handle the biochemical issues.

Reference:

1. Whitehead, K. A.; Langer, R.; Anderson, D. G. Knocking down barriers: advances in siRNA delivery. *Nat Rev Drug Discov* **2009,** *8* (2), 129-138.

2. Oh, Y. K.; Park, T. G. siRNA delivery systems for cancer treatment. *Adv Drug Deliver Rev* **2009,** *61* (10), 850-862.

3. Li, J. P.; Ouyang, Y. Y.; Kong, X.; Zhu, J. Y.; Lu, D. N.; Liu, Z. A multi-scale molecular dynamics simulation of PMAL facilitated delivery of siRNA. *Rsc Adv* **2015,** *5* (83), 68227-68233.

4. Paulsen, C. E.; Armache, J. P.; Gao, Y.; Cheng, Y. F.; Julius, D. Structure of the TRPA1 ion channel suggests regulatory mechanisms. *Nature* **2015,** *520* (7548), 511-+.

5. Semple, S. C.; Akinc, A.; Chen, J. X.; Sandhu, A. P.; Mui, B. L.; Cho, C. K.; Sah, D. W. Y.; Stebbing, D.; Crosley, E. J.; Yaworski, E.; Hafez, I. M.; Dorkin, J. R.; Qin, J.; Lam, K.; Rajeev, K. G.; Wong, K. F.; Jeffs, L. B.; Nechev, L.; Eisenhardt, M. L.; Jayaraman, M.; Kazem, M.; Maier, M. A.; Srinivasulu, M.; Weinstein, M. J.; Chen, Q. M.; Alvarez, R.; Barros, S. A.; De, S.; Klimuk, S. K.; Borland, T.; Kosovrasti, V.; Cantley, W. L.; Tam, Y. K.; Manoharan, M.; Ciufolini, M. A.; Tracy, M. A.; de Fougerolles, A.; MacLachlan, I.; Cullis, P. R.; Madden, T. D.; Hope, M. J. Rational design of cationic lipids for siRNA delivery. *Nat Biotechnol* **2010,** *28* (2), 172-U18.

6. Kanasty, R.; Dorkin, J. R.; Vegas, A.; Anderson, D. Delivery materials for siRNA therapeutics. *Nat Mater* **2013,** *12* (11), 967-977.

7. Meneksedag-Erol, D.; Tang, T.; Uludag, H. Molecular modeling of polynucleotide complexes. *Biomaterials* **2014,** *35* (25), 7068-7076.

462723 First Principles Computational Study on Adsorption of Organic Methyl Iodide in Triethylenediamine Impregnated Activated Carbon

Abstract:

First Principles Computational Study on Adsorption of Organic Methyl Iodide in Triethylenediamine Impregnated Activated Carbon

Hoje Chun[1], Byungchan Han[2]*

[1]*Integrated Science and Engineering Division, Yonsei University, Seoul, 03722, Republic of Korea*

[2]*Department of Chemical and Biomolecular Engineering, Yonsei University, Seoul, 03722, Republic of Korea*

We study removal of gas-phase organic iodine (CH_3I), which is one of the toxic chemical generated in nuclear power plant operation or accidents, via adsorption in triethylenediamine (TEDA) impregnated activated carbon (AC). We apply first principles density functional theory (DFT) calculations to understand how experimentally observed quaternary ammonium salts form in the surface. It is propose that the adsorption energy of a CH_3I shows substantial heterogeneity as function of varying the adsorption sites, porosity and H_2O environment. Most importantly, it is clearly shown by first principles calculated free energy diagrams for the adsorption processes that TEDA not only helps in reducing the activation barrier of the dissociation of CH_3I but also attracts H_2O molecules freeing the porous AC surface from poisoning by humidity and enhancing removal efficiency of the CH_3I. Our computational study enables to unveil detailed chemical process and thus, can open new ways to design highly functional AC in removing environmentally and biologically hazardous materials.

463583 Evaluation of Effective Stacking Interactions of Kaolinite Nanoparticles in Aqueous and Organic Solvents By Molecular Theory of Solvation

Abstract:

Crucial for the understanding of the rich phase behavior of nanocolloidal systems is accurate determination of effective interactions between the nanoparticles. Evaluation of the interactions by conventional molecular simulation techniques generally requires costly free energy integration and therefore is usually avoided by adopting simple and often unrealistic semiempirical effective potentials. In order to efficiently obtain effective interactions in a nanocolloidal system, we employ the three-dimensional reference interaction site model with the Kovalenko-Hirata closure relation (3D-RISM-KH) molecular theory of solvation [1-3]. 3D-RISM-KH molecular theory of solvation offers fully atomistic description with all-atom force-fields and notable performance advantages over molecular simulations. In addition, this modeling approach predicts the adsorption structure and thermodynamics of multicomponent systems and allows comparison of the interactions among solvent components with the adsorbent.

The approach is applied to study the effective interactions between kaolinite clay platelets, which serve as a simple model of Mature Fine Tailing (MFT). MFT is a gel-like colloidal suspension of clay nanoparticles and residual bitumen in water with electrolytes produced by aqueous process of extraction of bitumen from oil sands. Efficient dewatering and densification of the MFT are of great importance for decreasing the environmental footprint of the oil sands industry. Currently, efforts are undertaken to introduce a process of non-aqueous extraction of bitumen from oil sands, which would raise the energetic efficiency and avoid the water pollution of the conventional process. However, regardless of the extraction method, oil sands bitumen froth contains fines (clay particles coated with bitumen) above the 0.5% limit required for pipeline transportation.

Potential of mean force between kaolinite clay platelets with several mutual orientations as well as adsorption structure in different aqueous/organic solvents with additives/adjuvants is computed. The molecular insights provided by the 3D-RISM-KH molecular theory of solvation allow us to better understand the mechanisms of bitumen desorption and clay nanoparticle flocculation in aqueous and organic solvents, and should facilitate the rational design of adjuvants/flocculants specifically tailored for these purposes.

References

1. Kovalenko, A.; Hirata, F. *J. Chem. Phys.* **1999**, *110*, 10095-10112.
2. Kovalenko, A. In: *Molecular Theory of Solvation*; Hirata, F., Ed.; Kluwer Academic Publishers: Dordrecht, **2003**; pp. 169-275.
3. Kovalenko, A. *Pure Appl. Chem.* **2013**, *85*, 159-199.

464247 Mapped Averaging for Highly Efficient Evaluation of Fluid-Phase Properties By Molecular Simulation

Abstract:

"Mapped averaging" is a recently published scheme for the reformulation of ensemble averages.[1,2] The framework uses approximate results from statistical mechanical theory to derive new ensemble averages (mapped averages) that represent exactly the error in the theory. Calculation of the mapped averages by molecularâ€¨simulation can proceed without contamination by noise produced by behavior that has already been captured exactly by the theory. Consequently,â€¨accurate and precise values of properties can be obtained while using lessâ€¨computational effort, in favorable cases, many orders of magnitude less. The mapped-averaging framework rests on two key elements: (a) the choice of the mapping; and (b) the formulas yielding the free-energy derivatives (properties) once the mapping is specified. For crystalline systems, a harmonic approximation provides a suitable starting point, allowing simulation to compute precisely the anharmonic contribution to the properties. For fluid systems, the formulation of a good mapping is not as obvious. In this work we examine ways to formulate a mapping and present new ensemble-average formulas for properties of interest to fluid-phase systems. We demonstrate the performance of these methods on several properties for a variety of molecular models.

(1) Schultz, A. J.; Moustafa, S. G.; Lin, W.; Weinstein, S. J.; and Kofke, D. A. Reformulation of Ensemble Averages via Coordinate Mapping, *Journal of Chemical Theory and Computation* 2016.

(2) Moustafa, S. G.; Schultz, A. J.; and Kofke, D. A. Very fast averaging of thermal properties of crystals by molecular simulation, *Phys. Rev. E* 2015, *92*, 043303.

464068 Molecular Dynamics Simulation of Cations (K^+, Na^+, Rb^+, Mg^{2+} and Sr^{2+}) Adsorption at the Muscovite Mica - Aqueous Solution Interface

Abstract:

Structure, hydration and distribution of ions at mineral-aqueous solution interface varies significantly over the dilute to higher concentration of the aqueous solution. Ions emerges from dissociated electrolytes, competes with the hydrated ions of minerals at the interface. A molecular level understanding of the competitive adsorption between these ions at the interface provides valuable inputs to the fields like soil enrichment, chemical weathering and remediation of heavy metals. In order to understand the properties of these ions we studied adsorption isotherms of pure solutions of chlorides (NaCl, RbCl, $MgCl_2$ and $SrCl_2$) on muscovite mica surface at various concentrations. The surface K^+-ions are exchanged by the cations of aqueous solution depending upon their free energy. We observed that the adsorption position of cations K^+, Na^+, Rb^+, Mg^{2+} and Sr^{2+} on the muscovite mica surface are at 1.8, 1.76, 1.87, 2.35 and 2.82Å respectively, in good agreement with experimental results (Lee et al., 2013 and Park et al., 2006). The competitive adsorption between surface K^+- ions and respective cations of the aqueous solutions was reflected at the different regions of the isotherm profiles. Our free energy calculations are able to explain the domination of K^+- ions adsorption at the dilute concentrations and salting out phenomena as further increase in the concentration of the solutions. In addition to the adsorption isotherms and free energy calculations we evaluated the density distributions and the pair correlations to obtain better understanding.

464436 Understanding the Influence of Curcumin on Amyloid-β Aggregation at the Molecular Scale

Abstract:

Alzheimer's Disease (AD) persists as a threat to the lives of millions in the United States each year, ranking as the sixth leading cause of death among all ages. The disease progression is characterized by the development of highly damaging aggregates in the neuronal network of the brain due to protein misfolding. One of these malfunctioning proteins, Amyloid-β (Aβ), prompts the generation of plaques in the extracellular space of neurons caused by self-assembly of Aβ into insoluble fibrils. It is clear that the fibrils are highly toxic, however the extent of this toxicity continues to be explored. In addition, a number of groups are attempting to understand the aggregation process of Aβ to promote the discovery of chemical agents to interfere with fibril formation. Curcumin, a component of the spice turmeric, is one compound known to alter Aβ assembly. Although once thought to stop fibril formation completely, experimental evidence from studies by our group shows that this compound attenuates the formation of fibrillar Aβ thereby decreasing its toxicity. The molecular scale forces governing the interactions between curcumin and Aβ are hypothesized to stem from hydrophobic effects, but this mechanism is not entirely clear. Theoretical approaches to understanding this process provide a powerful means to investigate how curcumin binds to Aβ fibrils at the molecular scale. We are currently using classical Molecular Dynamics (MD) simulations to study this phenomenon at an atomistic scale with a focus on the critical binding sites and energetics that may be responsible. To this point, we have found multiple locations on short Aβ protofibrils that appear to attract curcumin. Complexation of two or more of these molecules looks to promote the affinity for Aβ as there are multiple locations where dimers of curcumin bind to fibrils after 20 ns.

464694 Computational Search for Magnetic Nanoparticles to Eradicate Cancer Cells Using the Hyperthermia Approach

Abstract:

In recent years, nanomaterials have received significant attention due to their unique properties. Magnetic nanoparticles (MNPs) in particular are of high importance due to their magnetic properties. These particles are being tested for various applications including cancer cell detection and targeted cell temperature increase.

Hyperthermia treatment of cancer cells has been one of the most promising ideas of last decade. The theory is to design materials which are very efficient in targeting the cancer cells. The designed particles are to have magnetic properties such that at certain magnetic field conditions the particle develops a higher temperature than the surrounding atmosphere/environment.

This project aims to investigate a way to model magnetic nanoparticle behavior in the presence of different magnitudes of magnetic fields. The aim is to model Brownian and Neel relaxation parameters to determine the heat generated by different MNPs to study which is best suited for hyperthermia treatment of cancer cells. For that, approaches were used to determine AC susceptibility of the out-of-phase particles which was then used in a 3-D model containing healthy and cancer cell. The 3-D model is used to determine temperature output for such materials in the presence of an optimal magnetic field.

Selected References:

1. Rudolf Hergt, Silvio Dutz, Robert Muller, Matthias Zeisberger, *Magnetic particle hyperthermia: nanoparticle magnetism and materias development for cancer treatment*, 2006, J. Phys.:Condens. Matter, Vol 18, S2919-S2934.
2. Shihu Wang and Elena E. Dormidontova, *Nanoparticle Design Optimization for Enhanced Targeting: Monte Carlo Simulations*, 2010, Biomacromlecules, Vol 11, 1785-1795
3. Andreas Jordan, Regina Scholz, Peter Wust, Horst Fahling, Ronald Felix, Magnetic fluid hyperthermia: *Cancer treatment with AC magnetic field induced excitation o biocompatible supermagnetic nanoparticles*, 1999, Journal of Magnetism and Magnetic Materials, Vol 201, 413-419
4. Stefan T Bromley, Iberio de P. R, Moreira, Konstantin M. Neyman, Francese Illas, *Approaching nanoscale oxides: models and theoretical methods*, 2009, Chem. Soc. Rev., Vol 38, 2657-2670.

464325 Computational Study of CO_2 Adsorption on Apohost and LiCl-Functionalized $Zn(bdc)(ted)_{0.5}$ Metal–Organic Frameworks

Abstract:

COMPUTATIONAL STUDY OF CO_2 ADSORPTION ON APOHOST AND LiCl-FUNCTIONALIZED $Zn(bdc)(ted)_{0.5}$METAL–ORGANIC FRAMEWORKS

Carlos Fernández-Cabán, Jorge Rosa-Raíces, Paul Meza-Morales, and María Curet-Arana

Department of Chemical Engineering, University of Puerto Rico, Mayaguez Campus, Road 108 - km 1.1, Mayaguez, PR 00681-9000, United States

Metal-organic frameworks (MOFs) are self-assembling crystalline nanoporous materials with potential applications in gas separation and storage and several other fields.[1–3] $Zn(bdc)(ted)_{0.5}$ (Zn-DMOF) is a MOF with remarkable uptake capacity and affinity for CO_2.[4,5] Studies involving Zn-DMOF have shown that its CO_2 adsorption properties can be enhanced by tailoring its structure and chemistry through post-synthesis modification.[6,7] An alternative functionalization method, known as spontaneous thermal dispersion, involves introducing extra-framework moieties that can adjust the material's functionality for a targeted application while preserving its basic chemical and structural properties. This method was recently used to incorporate LiCl in Zn-DMOF, and resulted in a crystalline functionalized variant (Zn-DMOF-(LiCl)) featuring a dramatically altered CO_2 uptake capacity, appearance of an adsorption–desorption hysteresis gap and contraction of the framework pores with respect to the apohost material.[8]

In this multiscale study, we use density functional theory (DFT) calculations and grand canonical Monte Carlo (GCMC) simulations to yield molecular-level insight into the structural and chemical features that underlie the CO_2 adsorption properties of Zn-DMOF and Zn-DMOF-(LiCl). Our computations reveal that the ionic salt residues in the functionalized variant remain bonded and interact preferably with the metallic centers of Zn-DMOF, inducing severe unit-cell deformation and altering the textural properties of the MOF as verified by comparing simulated and experimental spectroscopic data. GCMC simulations with the deformed structures are in fair agreement with experimental results over a significant portion of the pressure range studied, and confirm that the CO_2 uptake capacity of the apohost MOF undergoes drastic changes upon functionalization with LiCl, with framework-adsorbate interactions being strengthened at low pressures. We also explore the low-energy configuration space of Zn-DMOF as a function of CO_2 loading, and resolve the reported guest-induced structural changes in the apohost material.[9]Our work offers a molecular-level understanding of the structural and textural changes caused by spontaneous dispersion of an inorganic moiety in a prototypical MOF, the concomitant impact on its functionality, and the applicability of this technique in the rational design and tailoring of similar materials.

References:

(1) Li, J. R.; Ma, Y.; McCarthy, M. C.; Sculley, J.; Yu, J.; Jeong, H. K.; Balbuena, P. B.; Zhou, H. C. *Coord. Chem. Rev.* **2011**, *255*(15-16), 1791.

(2) Lee, J.; Farha, O. K.; Roberts, J.; Scheidt, K. A.; Nguyen, S. T.; Hupp, J. T. *Chem Soc Rev* **2009**, *38*(5), 1450.

(3) Kreno, L. E.; Leong, K.; Farha, O. K.; Allendorf, M.; Van Richard P., D.; Hupp, J. T. *Chem. Rev. (Washington, DC, United States)* **2012**, *112*, 1105.

(4) Mishra, P.; Mekala, S.; Dreisbach, F.; Mandal, B.; Gumma, S. *Sep. Purif. Technol.* **2012**, *94*, 124.

(5) Chaemchuen, S.; Zhou, K.; Kabir, N. A.; Chen, Y.; Ke, X.; Van Tendeloo, G.; Verpoort, F. *Microporous Mesoporous Mater.* **2015**, *201*, 277.

(6) Wang, Z.; Cohen, S. M. *J. Am. Chem. Soc.* **2009**, *131*(46), 16675.

(7) Cohen, S. M. *Chem. Rev.* **2012**, *112*(2), 970.

(8) Hernandez-Maldonado, a J.; Guerrero-Medina, J.; Arce-Gonzalez, V. C. *J. Mater. Chem. A* **2013**, *1*, 2343.

(9) Guerrero-Medina, J.; Mass-González, G.; Pacheco-Londoño, L.; Hernández-Rivera, S. P.; Fu, R.; Hernández-Maldonado, A. J. *Microporous Mesoporous Mater.* **2015**.

464514 Practical Data Sharing for Molecular Simulation

Abstract:

All molecular simulation occurs at a computer. Consequently, all of the procedures and results are at some point recorded on the computer in a machine readable format. Thoroughly documenting this information promises a multitude of benefits, not only to the community, but also to the author:

1. Eliminating reproducibility issues
2. Reducing time spent solving the same problems
3. Recording of own work organized by manuscript
4. Allowing for data to be used in other applications and accessed through data mining

Despite the potential advantages, such high-level data sharing remains uncommon. This is potentially due to the perceived difficulty of generating the required documentation.

Here we illustrate practical methods, already implemented in the Kitchin group[1], for streamlining the incorporation of high-level data sharing into the manuscript writing process. In this way, we hope to obtain the advantages listed above with little additional, or potentially even less effort than current methods.

1. Kitchin, John R. "Data Sharing in Surface Science." ScienceDirect, 16 May 2015.

465509 Multiscale Simulations of Protein Adsorption on Self-Assembled Monolayers

Abstract:

Protein adsorption plays an important role not only in a wide range of basic biological processes but also in many applications such as protein chromatography, drug delivery on solid substrates, biosensors, biofuel cells and biomaterials. The ability to tailor both head and tail groups of self-assembled monolayers (SAMs) makes them excellent systems for the fundamental understanding of protein adsorption. We investigated the adsorption behavior of various proteins (such as neuromedin-B, prototypical and mutated protein G B1, lysozyme, ribonuclease A, lipase, hydrophobin, cytochrome c, laccase, feruloyl esterase, and fibrinogen) on the SAMs terminated with different functional groups (such as methyl, hydroxyl, carboxyl, amino, sulfonate, sulfobetaine, phosphorylcholine, and mixed carboxyl/trimethylamino) under different conditions (such as pH, ionic strength (IS), ionic type, surface charge density (SCD), and external electric fields (EFs)) at the microscopic and mesoscopic scales using a combined parallel tempering Monte Carlo (PTMC) algorithm, coarse-grained molecular dynamics (CGMD), and all-atom molecular dynamics (AAMD) simulations. The results indicate that: (i) the conformation change of a protein on the hydrophobic surface is relatively bigger; (ii) the nonspecific protein adsorption can be resisted by the zwitterionic surface due to the tightly bounded water molecules; (iii) the electrostatic interactions between a protein and a charged surface are weakened when the ionic strength increases; (iv) a counterion layer will be formed near the surface when the SCD and IS are high, which can mediate the adsorption of a protein with the same sign of net charge; (v) the orientation of a protein adsorbed on a charged/neutral surface is dominated by its electric/hydrophobic dipole; (vi) the external electric fields can modulate protein adsorption behaviors on the charged surfaces. Our work might provide some guidance for the design of novel ligands for protein purification, design of medical materials with better biocompatibility, improving the efficiency and activity of the immobilized enzymes, the design of biosensors with better responsiveness, and design of antifouling materials with stronger ability in resisting protein adsorption.

465351 Sour Gas Sweetening By Adsorption in Zeolites

Abstract:

Natural gas contributes a third of the fossil fuel usage in the United States, and with the discovery of new gas fields, this fraction is on the rise. Hydrogen sulfide removal is a long-standing economic and environmental challenge facing the natural gas industry. Our goal is to develop next-generation solutions that efficiently capture hydrogen sulfide from natural gas mixtures. To this end, we are exploring separation processes using zeolites – nanoporous materials with highly specific pore sizes and shapes. H_2S is a highly toxic gas, and performing a wide experimental screening of all zeolites will not only be a very expensive endeavor, but will also require impeccable safety measures. This is an apt situation for molecular modeling to take the lead and guide *real* experiments.

We present here the development of a new hydrogen sulfide force field that accurately describes a wide set of physical properties such as the pure-component and binary vapor–liquid equilibria with CH_4 and CO_2, the liquid and solid structures, the relative permittivity, the triple point, and the gas-phase diffusion behavior.[1] Using this new H_2S model, we have computationally investigated separation using various zeolitic frameworks and have identified promising sorbents for gas sweetening.[2,3]

[1] M. S. Shah, M. Tsapatsis, and J. I. Siepmann, 'Development of the transferable potentials for phase equilibria model for hydrogen sulfide,' *J. Phys. Chem. B* **2015**, 119, 7041–7052.

[2] M. S. Shah, M. Tsapatsis, and J. I. Siepmann, 'Monte Carlo Simulations Probing the Adsorptive Separation of Hydrogen Sulfide/Methane Mixtures using All-silica Zeolites,' *Langmuir* **2015**, 31, 12268–12278.

[3] M. S. Shah, M. Tsapatsis, and J. I. Siepmann, 'Identifying Optimal Zeolitic Sorbents for Sweetening of Highly Sour Natural Gas,' *Angew. Chem. Intl. Ed.* **2016**, 55, 5938–5942.

466569 Effects of Ionic Liquids on HP35 Folding Thermodynamics

Abstract:

Ionic liquids (ILs) are a class of solvents comprised of salts that are liquid below 100°C. ILs have been shown to affect the structure and function of many proteins and enzymes. The magnitude and nature of these effects varies from protein to protein and from IL to IL. Molecular simulations can be employed to better understand the ways in which ILs interact with proteins. In this study, we measure the effects of IL-water mixtures on the the folding thermodynamics of a model protein (HP35) using microcalorimetry experiments and simulations utilizing parallel tempering metadynamics in the well-tempered ensemble. These simulations quickly elucidate the free energy and entropy of unfolding for HP35. Furthermore, the results of the simulations can be reweighed to project the free energy onto any collective variable of choice. These methods can be used to better understand detailed interactions of a protein with ILs or to reparametrize force fields to more accurately reproduce thermodynamic observables.

467564 Sintering of Bimetallic Ag-Au Nanoparticles By Molecular Dynamics

Abstract:

Abstract

Nanoparticles have attracted strong interest due to their novel properties and have been explored widely in photonics, electronics, catalysis, sensors, information storage, photography, imaging and biological labeling. For example, gold nanoparticles are used extensively in catalysis and silver nanoparticles have remarkable antibacterial properties. However, the former exhibit reduced electron transfer on Au(111) surface, hindering the adsorption of O_2 on such surfaces, while the latter are highly cytotoxic prohibiting their use in biomedical applications. Bimetallic nanoparticles (e.g. Ag-Au) often exhibit enhanced chemical, catalytic and plasmonic properties compared to their corresponding parent materials and, thus, can enhance the affinity with O_2compared to pure gold nano-catalysts or decrease the cytotoxicity of silver (Sotiriou et al., 2014).

Physical properties (e.g. reactivity and ligand adsorption) of such nanostructures depend on the particles crystal structure and surface facet orientation. Experimental investigation of particlesâ€™ nanocrystallinity with electron microscopy leads to incomplete characterization as it is based on the analysis of 2D projections of nanoparticles. However, Molecular Dynamics (MD) simulations can be used to provide physical insight in crystal structure changes and dynamics with atomistic detail (Buesser and Pratsinis, 2015) and can complement the above experiments, especially for very small nanoparticles.

Here, MD simulations are used to systematically investigate the sintering mechanism and crystallinity dynamics of Ag-Au nanoparticles of various sizes and temperatures. The sintering rate of coalescing Ag-Au nanoparticles of various

initial morphologies (segregated, core-shell or alloys) is investigated for different particle sizes and temperatures. The characteristic sintering time of Ag-Au nanoparticles (determined by tracking the evolution of the surface area during sintering or coalescence) can be bracketed in-between those of pure Ag and Au nanoparticles.

Furthermore, particle crystallinity depends on the synthesis process conditions (Goudeli and Pratsinis, 2016) and lead to the exposure of different facets affecting surface chemistry and product particle morphology and their catalytic properties (e.g. catalytic activity and selectivity). The atoms at the grain or particle boundaries are disordered compared to the bulk atoms, especially during particle adhesion. Such defects (e.g. grain boundaries) at the atomic scale can strongly affect electrical, optical and mechanical properties of 2D materials.

Keywords:sintering rate, bimetallic silver-gold, molecular dynamics.

References

Buesser, B., and Pratsinis, S.E. (2015) *J. Phys. Chem. C*, 119, 10116-10122.

Goudeli, E., Eggersdorfer, M.L., and Pratsinis, S.E. (2015) *Langmuir*, 31, 1320-1327.

Goudeli, E., and Pratsinis, S.E. (2016) *AIChE J*, 62, 589-598.

Sotiriou G A, Etterlin G D, Spyrogianni A, Krumeich F, Leroux J-C, and Pratsinis S E. (2014). *Chem. Commun.* 50: 13559-13562.

466426 Fluid Transport through Nano-Pores: A Stochastic Approach

Abstract:

Fluid transport through nano-pores: a stochastic approach

Maria Apostolopoulou [a], Alberto Striolo [a†], Michail Stamatakis [a†], Richard Day [b]

[a] University College London, Department of Chemical Engineering, Torrington Place, London WC1E 7JE, United Kingdom.

[b] Halliburton, Building 4 Chiswick Park 566, Chiswick High Road, London W4 5YE, United Kingdom.

†Corresponding authors: a.striolo@ucl.ac.uk, m.stamatakis@ucl.ac.uk

Keywords: computational molecular engineering, fluid mechanics, shale gas

Abstract

The mechanism of fluid migration in porous networks continues to attract great interest. Although single and two-phase flow systems have been successfully studied in macroscale and microscale channels, the transport of fluids through networks of nanochannels is yet to be fully understood. The Darcy's law (phenomenological continuum theory) is often used to describe macroscopically fluid flow through a porous material and can efficiently describe the fluid transport through micro pores, in which the dominant regime is viscous flow. However, it appears that Darcy's formalism fails to describe flow in nano-channels, in which some studies have suggested that water and air exhibit several orders of magnitude higher flowrate than that predicted by continuum theory. This observation, first reported by Klinkenberg in 1941, was attributed to the slip-flow effect.

Fluid transport can be studied by using several computational techniques, such as Molecular Dynamics or Lattice-Boltzmann, each with its own limitations. Molecular Dynamics simulations are used to describe the trajectories of individual atoms and allow us to follow the dynamics of molecular processes in great detail. However, they are incapable of probing large systems for long times, due to computational limitations. Lattice-Boltzmann simulations can provide larger time scale analysis, while accounting the slip boundary condition, at the expense of molecular-level phenomena that might become important in nanopores.

To overcome the current limitations, we implement a stochastic approach to describe fluid transport through porous networks. This method is based on the multi-variate master equation, by which the space is represented as set of connected finite volumes (voxels), and transport is simulated as a random walk of molecules, which hop from voxel to voxel with a given propensity. We validated the approach using a 1D model consisting of two neighbouring voxels. The results were compared against deterministic equations that describe the system at equilibrium. The key parameters for the stochastic approach were the energy barrier between the two states (voxel occupancies) and the diffusion coefficients. We examined the equilibrium occupancies of voxels and the error between the stochastic and the deterministic approach for different populations of molecules (100, 1000, 10000 and 100000) while changing the initial configuration. While the stochastic approach yields errors of less than 1%, the lack of limitations on occupancies could be a source of uncertainty in other scenarios. Overall, these results indicate that the model can be extended to accurately describe fluid transport in 1D or 2D networked systems, in which single phase or multiphase flow takes place. We will present here results obtained for models representing simplified versions of shale rocks.

467709 Structural Changes on Coordination Polymer Ligands (CPL-bpp) Induced By CO_2: Theoretical Study Based on Density Functional Theory and Grand Canonical Monte Carlo Simulation

Abstract:

Coordination pillared-layers (CPLs) are nanoporous materials that consist of layers formed by copper and ionic pyrazine-2,3-dicarboxylate, and they are separated by nitrogenated organic linkers. It has been previously demonstrated that CPLs could potentially be exploited for CO_2 capture applications.[1–3] Some of these materials, such as CPL-2, CPL-4, and CPL-5, exhibit structural changes triggered by CO_2 adsorption, which results in an enhancement of the pore volume leading to an increment of CO_2 adsorption uptake.[2,4,5]

CPL-bpp is a member of the CPL series, which consists of layers formed by copper and ionic pyrazine-2,3-dicarboxylate that are separated by 1,3-bis(4-pyridyl)propane (*bpp*). This material possesses uniform pore channels, and it is a promising material for CO_2 air-capture applications because it has remarkable adsorption affinity for CO_2 compared to other molecules, such as O_2, N_2, and CH_4.[6]

In this work, we used periodic density functional theory (DFT) calculations to elucidate the structural changes induced by CO_2 on this material. Calculations were performed using PBE as exchange-correlation functional in VASP. We optimized the unit cell of CPL-bpp with various CO_2 molecules in the system to quantify the effect of CO_2 on the unit cell parameters. Our results indicate that CO_2 induces changes in the unit cell of CPL-bpp, and the magnitude of these changes are dependent on the number of CO_2 molecules in the system. Our results also demonstrate that the aromatic rings on the *bpp* ligands rotate with increasing number of CO_2 molecules. Grand canonical Monte Carlo (GCMC) simulations were performed to

obtain CO_2 simulated isotherms, which were compared with experimental data available in the literature. Our results shed light into the interplay between CO_2 adsorption and CPL-bpp structural changes triggered by its interaction with CO_2.

REFERENCES

(1) García-Ricard, O. J.; Silva-Martínez, J. C.; Hernández-Maldonado, A. J. Systematic Evaluation of Textural Properties, Activation Temperature and Gas Uptake of $Cu_2(pzdc)_2L$ [L = Dipyridyl-Based Ligands] Porous Coordination Pillared-Layer Networks. *Dalt. Trans.* **2012**, *41*, 8922–8930.

(2) Riascos-Rodríguez, K.; Schroeder, A. J.; Arend, M. R.; Evans, P. G.; Hernández-Maldonado, A. J. Hysteretic Adsorption of CO2 onto a $Cu_2(pzdc)_2(bpy)$ Porous Coordination Polymer and Concomitant Framework Distortion. *Dalton Trans.* **2014**, *43*, 10877–10884.

(3) Chen, H.; Riascos-Rodríguez, K.; Marcano-González, M. E.; Hernández-Maldonado, A. J. $Cu_2(pzdc)_2L$ [L=dipyridyl-Based Ligands] Porous Coordination Polymers: Hysteretic Adsorption and Diffusion Kinetics of CO_2 and CH_4. *Chem. Eng. J.* **2016**, *283*, 806–815.

(4) Matsuda, R.; Kitaura, R.; Kubota, Y.; Kobayashi, T. C.; Takata, M.; Kitagawa, S. Incommensurate Guest Adsorption in Bellows-Shaped One-Dimensional Channels of Porous Coordination Polymers. *Microporous Mesoporous Mater.* **2010**, *129*, 296–303.

(5) Meza-Morales, P. J.; Gomez-Gualdron, D.; Curet-Arana, M. C. Insights into the Coordination Polymer Ligand Deformation upon CO_2 Adsorption through Grand Canonical Monte Carlo Simulations. In *AiChe Annual Meeting*; Salt Lake City, UT, 2015.

(6) Hernández-Maldonado, A. J.; Arrieta-Pérez, R. R.; Primera-Pedrozo, J. N.; Exley, J. Structure of a Porous $Cu_2(pzdc)_2(bpp)$ (Pzdc: Pyrazine-2,3-Dicarboxylate, Bpp: 1,3-Bis(4-Pyridyl)propane) Coordination Polymer and Flexibility upon Concomitant Hysteretic CO_2 Adsorption. *Cryst. Growth Des.* **2015**, *15*, 4123–4131.

468087 Recent Advances in the Development of Cassandra: An Open Source Monte Carlo Framework for Phase Equilibria Calculations

Abstract:

The chemical sciences rely on a set of fundamental mathematical theories that serve as a basis to various computational modeling techniques. These methodologies have proven to be powerful tools to understand matter at the microscopic level. Among these techniques, quantum mechanical (QM) calculations provide the most fundamental description of matter by numerically solving the Schrödinger wave equation. In principle, any chemical system can be described by this equation. However, obtaining its solution is a formidable task even for relatively simple systems. Consequently, approximations must be made to address problems whose length and time scales are inaccessible by QM methods. One path for working toward this objective is classical molecular simulations. These techniques approximate atomic interactions using continuous interaction potentials, which can be developed empirically or through quantum theory. Within classical simulations, two main set of methodologies are used: molecular dynamics (MD) and Monte Carlo (MC). The principle of the first method lies on the numerical solution of Newton's equations of motion to propagate the system, whereas the second stochastically generates states of a system according to a predetermined statistical mechanical probability distribution.

Our group has developed Cassandra, an open source MC framework that allow the simulation of complex chemical systems within different statistical mechanical ensembles. Cassandra has been used to simulate fluids that involve a wide range of molecules such as water, alcohols, carbon dioxide, ionic liquids or mixtures of these. In this poster, recent advances in the development of this software package will be described. New capabilities include the implementation of advanced methodologies based on expanded ensembles to study dense systems, incorporation of polarization through fluctuating charges and Drude oscillators, implementation of non-Boltzmann sampling methodologies, the addition of fast pairwise alternatives to Ewald summation, the ability to simulate adsorption in porous materials and the implementation of neighbor lists. In addition, a series of scripts and utilities have been developed to help the user set up calculations faster and with fewer errors.

467583 A Transferable Multi-Resolution Coarse-Grained Model for Amorphous Silica Nanoparticles

Abstract:

Silica nanoparticles are ubiquitous in modern day nanoscience, with drug delivery, biomedical imaging, and polymer nanocomposites highlighting a host of pertinent applications. The balance between attractive and repulsive forces at the atomic level governs assembly of these materials into complex macroscale structures with a wide range of properties. Molecular dynamics simulations provide an avenue to examine nanoparticles with atomic-level resolution and to precisely tune the interactions between them. However, relevant nanoparticle sizes typically feature tens or hundreds of thousands of atoms, making simulations of more than a few nanoparticles computationally burdensome. Several coarse-grained (CG) models have been developed for amorphous silica nanoparticles in attempt to overcome this challenge, e.g., Ghanbari et al. constructed nanoparticles using CG particles representing individual SiO2 subunits [1] and Lee and Hua developed a model where CG beads represented spherical Si6O12 subunits [2]. However, neither of these models were compared directly to experimental or all-atom simulation data for amorphous silica nanoparticles and thus their efficacy and accuracy remain unproven.

In this work, we develop a CG model for amorphous silica nanoparticles using a potential-matching approach. Here, nanoparticles are represented as a collection of individual CG beads, modeled via a Mie potential, whose parameters are tuned such that the effective interaction of the collection of CG particles matches the interaction between atomistically detailed amorphous silica nanoparticles. To generate potentials that are transferable, CG parameters are derived to simultaneously match a range of nanoparticle sizes. Collective optimization of the parameters makes use of the constrained optimization by linear approximation (COBYLA) method provided by the openMDAO framework [3], resulting in CG potentials that match their atomistic counterparts with high fidelity. Higher and lower resolution models are also presented, corresponding to changes in the CG particle surface density, where such changes are made trivially using the mBuild software package [4]. Point particle models of NPs, representing the lowest resolution of coarse-graining, are also presented. We further demonstrate the extension of this potential matching procedure to obtain cross-interactions between CG particles and alkanes, allowing alkane-grafted CG amorphous silica nanoparticles to be constructed, whose properties show close agreement with atomistic models.

[1] A. Ghanbari, T. V. M. Ndoro, F. Leroy, M. Rahimi, M. C. Böhm, and F. Müller-Plathe, "Interphase Structure in Silica–Polystyrene Nanocomposites: A Coarse-Grained Molecular Dynamics Study," Macromolecules, vol. 45, no. 1, pp. 572–584, 2012.

[2] C. K. Lee and C. C. Hua, "Nanoparticle interaction potentials constructed by multiscale computation," J. Chem. Phys., vol. 132, no. 22, 2010.

[3] J. Gray, K. T. Moore, T. A. Hearn, and B. A. Naylor, "Standard Platform for Benchmarking Multidisciplinary Design Analysis and Optimization Architectures," AIAA Journal, vol. 51, pp. 2380-2394, 2013.

[4] C. Klein, J. Sallai, T. J. Jones, C. R. Iacovella, C. McCabe, P. T. Cummings, "A hierarchical, component based approach to screening properties of soft matter," Accepted at Mol. Mod. and Sim.: Applications and Perspectives, 2016.

468042 Elucidating the Molecular Architecture of the Blood-Brain Barrier Tight Junctions

Abstract:

The blood-brain barrier is a selective permeability molecular barrier constituted by claudin-5 tight junction strands. Exploring treatment and therapeutics options for neurodegenerative diseases such as Alzheimerâ€™s and Parkinsonâ€™s remains challenging because of limited understanding of the claudin-5 tight junction permeability. In an effort to overcome this hurdle, we use multiscale molecular simulations to study the intricate self-assembly of claudin-5 transmembrane proteins in explicit membrane and solvent environment. Based on the simulation we were able to identify the stable dimer conformations that result from membrane driven claudin-5 *cis* interactions and the corresponding pore structures. Our results reveal the nature of the molecular interactions that drive claudin-5 aggregation in forming the tight junctions, and are in good agreement with the previous experimental predictions. The insights form this study will help in designing the next generation of smart therapeutics that can permeate the blood-brain barrier.

468528 Computational Analysis of the Assembly and Catalytic Effects of Self-Assembled Monolayers

Abstract:

Self-assembled monolayers (SAMs) can be used to enhance selectivity in various nanoparticle catalyzed reactions by altering the availability of certain active sites or interacting with surface reaction species. We use density functional theory (DFT) to investigate the adsorption behavior of alkane-thiolates over Pd step and terrace surfaces. We also investigate reaction pathways for various hydrogenation reactions on SAM coated/uncoated surfaces, and compare results with experimental reactivity data. DFT calculations indicate that, at lower coverages, small chain alkane-thiolates prefer to adsorb on under-coordinated step sites. Due to stronger van der Waals interaction in longer chain thiolates, at higher coverage, long chain alkane-thiolates form highly ordered self-assembled structures on the terrace sites. This leaves the step sites open for catalytic reaction, selectively facilitating specific reaction paths. One example of such a selective path is the hydrogenation of furfural to furfuryl alcohol. On an uncoated Pd surface, the furan ring in furfural adsorbs lying flat on the terrace sites, facilitating decarbonylation to form furan. However, on SAM coated Pd catalysts, the furfural to furan pathways are hindered due to the SAM occupying the terrace sites. This forces the furan ring in furfural to attach upright on the step sites via the carbonyl oxygen, facilitating the conversion pathway to furfuryl alcohol. DFT results examining thiol assembly on different Pd facets, and its dependence on the chemical potential will also be discussed.

468701 A Software Ecosystem for the Data-Driven Design of Chemical Systems and the Exploration of Chemical Space

Abstract:

Trial-and-error research approaches are increasingly ill equipped to meeting the complex challenges involved in the discovery and design of next-generation chemistry and materials. Our work recognizes the great opportunities that are arising with the shift towards data-driven *in silico* research and a rational design paradigm. These approaches are poised to mitigate the inefficiencies, shortcomings, and limitations of traditional trial-and-error research. However, the notion to utilize modern data science in the chemistry context is so recent that much of the basic infrastructure has not yet been developed, or is still in its infancy. The existing tools and expertise tend to be in-house, specialized, or otherwise unavailable to the community at large. Data science is thus in practice beyond the scope and reach of most researchers in the field. Our work aims to chart new paths in this area by creating an open, general-purpose software ecosystem designed to overcoming this situation, filling the prevalent infrastructure gap, and thus making data-driven research a viable and widely accessible proposition. Our software ecosystem fuses *in silico* modeling (in particular computational quantum chemistry), high-throughput screening techniques, and Big Data analytics into an integrated research infrastructure. We have been developing the necessary methods, algorithms, protocols, and codes, and assembled them in three loosely connected program suites: ***ChemHTPS*** provides an automated platform for the virtual high-throughput screening of compound and material candidate libraries as well as reaction networks; ***ChemBDDB*** offers a database and data model template for the massive information volumes created by data-intensive projects; and ***ChemML*** is a machine learning and informatics toolbox for the validation, analysis, mining, and modeling of such data sets.

468893 Adaptive Coarse-Graining of Molecular Dynamics through Diffusion Maps

Abstract:

In many molecular dynamics (MD) applications, the system variables (or their statistics) quickly become slaved to a reduced set of coarse variables, which parameterize a low-dimensional manifold in the high-dimensional space of the full system. However, exploration of the energy landscape on the manifold can be hindered by deep potential wells and high transition barriers. Methods to alleviate this time limitation, such as umbrella sampling and equation-free modeling, require advance knowledge of good coarse variables.

When this advance knowledge is not available, we couple equation-free methods to diffusion maps, a nonlinear manifold learning algorithm, to learn the coarse variables â€œon the flyâ€ based on relatively short bursts of the MD simulator. By establishing the low dimensionality of an attracting manifold and geometrically extrapolating outwards, the simulator can be biased towards rare events, effectively moving up potential gradients and out of potential wells. This allows the simulator to time-effectively search the potential landscape for new stable states and transition pathways. We demonstrate our approach on a toy stochastic simulator as well as MD simulations of alanine dipeptide.

469346 Molecular Dynamics Simulation of Water and Ion Transport through Carbon Nanotubes

Abstract:

Carbon nanotubes possess smooth hydrophobic surfaces and can be chemically functionalized enabling advanced separation processes by permitting high water flow rates while efficiently partitioning ions. Application of continuum theories to describe transport of water and ions may become difficult at the nanoscale and thus alternate techniques that help to evaluate structural and dynamical features in atomic detail such as molecular dynamics simulations are often used. We present our results from molecular dynamics simulations of pure water with two different water models (SPC/E and TIP4P/2005) and a 1:1 aqueous solution of NaCl confined in sub-3 nm diameter finite carbon nanotubes. The equilibrium density profiles of water and ions in these narrow pores are shown and their correlations with the observed dynamic properties in terms of diffusion coefficients obtained via the Einstein equation and the Green-Kubo relation are discussed. The transport properties are further evaluated by calculation of the permeation events of water and ions through charged and uncharged nanotubes. Additionally, we also obtain measurements of ion occupancy in the nanotube, hydration numbers of ions, water and ion fluxes, and ion rejection rates. The role of confinement and charge on the physical features of water and ion transport through carbon nanotubes obtained via MD simulation will be presented in this poster.

469367 Understanding the Structure and Function of Enzymes in ILs for Improved Biocatalysis

Abstract:

Evaluating the degree to which an enzyme is able to maintain its structure and function in non-native solvents has widespread relevance, particularly to the field of biocatalysis. Ionic liquids (ILs) represent one such category of non-native solvents that have been gaining ground in recent decades as important solvents for biocatalysis, due to their low flammabilities and vapor pressures and highly tunable solvent properties. The discovery that some enzymes are able to maintain a level of function/activity in ILs has led to large-scale interest in determining the extent of the broadened landscape of viable biotransformations. Despite this interest, however, progress has been hindered due to a lack of a molecularly detailed understanding of how to rationally design enzymes to have increased stability in ILs. Molecular simulation has proven to be an invaluable tool at providing insight into the atomic level interactions of enzymes with ILs. Included in this poster are results from molecular dynamics (MD) simulations of multiple different enzyme/IL systems of interest: (1) a newly discovered family of enzymes called lytic polysaccharide monooxygenases (LPMOs) are simulated in three aqueous ILs at 0, 10, and 20 weight percent to evaluate their potential to aid cellulases in the decomposition of cellulose for conversion into fuels and chemicals, and (2) *Bacillus subtilis* lipase A is simulated in aqueous [BMIM][Cl] at 0, 5, 10, and 20 volume percent to evaluate its potential to maintain stability and activity in ILs for general biocatalysis applications. Structural and dynamic analyses indicate LPMO enzymes are predicted to be highly stable across the range of IL solutions tested; however, key areas that deviated from the crystal structures are identified as a basis for future rational design efforts to increase stability in ILs through surface charge modifications. Additionally, preferred binding sites for IL ions on the surface of lipase A are identified and compared to experimental crystallographic data[1] as a means to stabilize against IL-induced inactivation.

References

[1] Nordwald, E.; Plaks, J.; Snell, J.; Sousa, M.; Kaar, J. Crystallographic Investigation of Imidazolium Ionic Liquid Effects on Enzyme Structure. *Chembiochem* **2015**, *16*, 2456-2459.

470176 A General Tool Chain for Screening of Soft Materials

Abstract:

We present a set of tools [1] used to generate arbitrary starting configurations for molecular dynamics simulations and enable users to rapidly screen the parameter space of soft materials with a particular focus on systems involving functionalized surfaces. The core tool, mBuild [2], generates starting configurations by minimizing or even eliminating the need to explicitly translate and orient components when building systems - users simply state which components to connect. mBuild supports parameterized structures through generative modeling. This allows for declaratively expressing repetitive structures, e.g. polymer chains, crystal structures, planar or spatial tiling, as well as for parameterizing the affine transformations applied to subcomponents of a composite component. For end users, this minimizes and often eliminates the need to explicitly rotate and translate components when assembling systems. Additionally, the approach enables users to programmatically vary parameters for a family of systems (e.g., polymer chain length) or interchange individual components (e.g., polymer type), while still employing the same general framework. Building upon prior work [2], we demonstrate the utility and flexibility of our tool chain by examining the frictional properties of several families of monolayers. Specifically, we explore the parameter space of alkane and PEG monolayers on both crystalline and amorphous substrates with varying surface patterns. Surface density and patterning of the monomers as well as monomer chain length are trivially tunable via the mBuild python interface that is designed to integrate with the scientific Python stack.Â As such, this allows the rapid generation of unique input ï¬les, which were used as input to perform parameter sweeps of the structural and tribological behavior of the monolayers.

[1] J. Sallai, G. Varga, S. Toth, C.R. Iacovella, C. Klein, C. McCabe, AÌ. LeÌdeczi, P.T. Cummings. "Web- and Cloud-based Software Infrastructure for Materials Design." The 14th International Conference on Computational Science (ICCS), Procedia Comput. Sci., 2014, 29, 2034-2044

[2] C. Klein, J. Sallai, T.J. Jones, C.R. Iacovella, C. McCabe, P.T. Cummings. "A hierarchical component based

approach to screening properties of soft matter." accepted at Mol. Mod. and Sim.: Applications and Perspectives

470185 Integrating Molecular Modelling in Systems Engineering

Abstract:

The Molecular Systems Engineering team at Imperial College has championed the close integration of molecular modelling towards the intelligent molecular design of materials for engineering applications. A workhorse of the group is the development of the newest version of the Statistical Associating Fluid Theory equation of state (SAFT-γ) [1-3], that is effectively used to bridge the scales from molecular modelling to process and product design [4,5]. The SAFT model employs intermolecular potential parameters that are estimated directly from macroscopic thermophysical properties (e.g., fluid-phase equilibria) using the equation of state. This procedure provides a robust route to obtain coarse grained force fields which can then be used to obtain transport and interfacial properties [6-10].

Examples are provided in the areas of modelling of non-ionic surfactant behaviour, (direct calculation of critical micelle concentrations of C_iE_j alkyl-ethoxylates, superspreading of siloxane-based surfactants, interfacial tensions of light switching surfactants, liquid-liquid separations of polystyrene-solvent mixtures) where coarse grained models are used in a group-contribution fashion to obtain phase behaviour and molecular structure and self-assembly properties.

The presentation furthers discusses the "bottled SAFT" approach [11] where as the parameters for coarse grained moieties can be approximated through a corresponding states formulation of SAFT-γ. The resources in Bottled SAFT allow not only the fetch of the parameters but also the immediate production of scripts to be used in common MD suites. It currently hosts parameters for over 5000 molecules.

The SAFT model exemplifies a systematic and universal way to obtain transferable and representative potentials which can be employed to make molecular modelling directly accessible to non-experts by removing the barriers to developing intermolecular potentials.

References

[1] V Papaioannou, T Lafitte, C Aveñdano, CS Adjiman, G Jackson, EA Müller, and A Galindo, Journal of Chemical Physics, 140, 054107 (2014)

[2] T Lafitte, A Apostolakou, C Avendaño, A Galindo, CS Adjiman, EA Müller, and G Jackson, Journal of Chemical Physics, 139, 154504 (2013)

[3] S Dufal, T Lafitte, AJ Haslam, A Galindo, GNI Clark, C Vega, G Jackson, Molecular Physics 113, 948-984 (2015).

[4] J Burger, V Papaioannou, S Gopinath, G Jackson, A Galindo, CS Adjiman, AIChE Journal 61, 3249-3269 (2015)

[5] CS Adjiman, A Galindo, G Jackson, Computer Aided Chemical Engineering 34, 55 (2014)

11154-11169 (2011)

2733 (2013)

[8] T Lafitte, C Avendaño, V Papaioannou, A Galindo, CS Adjiman, G Jackson, EA Muller, Molecular Physics 110, 1189-1203 (2012)

[9] O Lobanova, C Avendaño, T Lafitte, EA Müller, G Jackson, Molecular Physics 113, 1228-1249 (2015)

[10] EA Müller, G Jackson, Annual Review of Chemical and Biomolecular Engineering 5, 405-427 (2014)

[11] www.bottledsaft.org

469538 Evaluating Force Fields Using Wolf$_2$pack's Molecular Database

Abstract:

Force fields are a fundamental element in molecular mechanics (MM) based methods, such as molecular dynamics (MD) simulations. Being able to quantitatively assess existing force-field, reoptimize parameters found to be questionable, and optimizing new parameters is critical for performing reliable MD simulations and in evaluating the resulting data. We have developed Wolf$_2$Pack (W_2P) that allows us to do these things. W_2P is a collection of scripts that allows us to efficiently generate quantum mechanic target observables (i.e. relative energies, geometries, electrostatic potentials) and generate their corresponding MM observables for direct comparison. An important component of W_2P is its molecular database that allows us to explore diverse chemical motifs and enables a high degree of reproducibility and transparency in the raw data. In this talk I will discuss how we use W_2P and its database to evaluate exiting force fields in order for users to better understand their strength and possible weaknesses. For example, we will show the degree of dependencies that MM relative potential energies have upon the choice of partial atomic charges (e.g. AMBER's RESP versus AM1-BCC, Glycam06's charges for force-field development versus those used its database of carbohydrate residues.)

470925 *Nusimm*: A User-Friendly Cloud Based Web Interface for Performing Molecular Simulations

Abstract:

Developing cloud based applications can eliminate many issues developers can run into when deploying software to their end users. Requiring users to install software package dependencies can often deter or inhibit users from utilizing a potentially useful application that could significantly advance their research. Centralizing an application in the cloud removes the burden from end users to configure the computing environment required for the software to work efficiently. This successfully expands the potential user base for an application to more people including those that previously would not have attempted to use the application due to a lack of expertise in complex computational science techniques.
We have developed *nusimm*, a web application powered by the open-source molecular simulation platform *pysimm* and built on the NSF funded cloud computing platform provided by nanoHUB which was designed to make performing molecular simulations easier. Cloud computing tools on nanoHUB can be accessed using any modern web browser. Input for simulation tools are collected through a graphical user interface in your browser and sent to computers in the cloud which perform simulation, collect results, and display results back in your web browser with the option of downloading data to analyze on your own machine. Simulation sessions can also be shared between users to help facilitate collaboration and educate new researchers about simulation techniques without requiring users to have extensive knowledge about high performance computing.
nusimm has been used both at the undergraduate and graduate level in courses aimed at introducing students to molecular simulations. Students with little knowledge of computational science were able to learn about two different types of polymeric systems and compare structural properties such as N_2 accessible surface area and pore size distributions. However *nusimm* provides not only teaching opportunities but also new avenues for research, collaboration with experimentalists, as well as a method to transparently demonstrate the reproducibility of previous work published in literature.

471242 Elucidation of Microstructure and Its Connection to Thermophysical Properties of Binary Ionic Liquid Mixtures, a Molecular Simulation Study

Abstract:

Ionic liquids (ILs) are molten salts composed entirely of ions and have inherent wide chemical variety. The property behavior of these so called designer solvents can be appropriately tuned through proper ion selection or use of ionic liquid mixtures. An array of intermolecular interactions is known to govern the structure of ILs which affects its physical and chemical properties. For pure ionic liquids, a thorough understanding of structure-property relationship has become possible from experimental and molecular simulation studies conducted over the last 15 years. However, such relationships are only now beginning to be probed for binary ionic liquid mixtures. Using atomistic molecular dynamics simulations, a study to obtain the relation between different interactions and physical property trend of ionic liquid mixtures was carried out. The work focused on two different mixture systems containing common cation 1-*n*-butyl-3-methylimidazolium $[C_4mim]^+$ in combination with chloride Cl^-Â and methylsulfate $[MeSO_4]^-$ and Cl^-Â and bis(trifluoromethanesulfonyl)imide $[NTf_2]^-$â€€ anions. Thermodynamic properties such as densities and excess molar volumes were computed. The binary mixture system $[C_4mim]Cl[MeSO_4]$ showed negative excess molar volumes in contrast to positive excess volumes displayed by $[C_4mim]Cl[NTf_2]$. Cation-anion and cation-cation interactions were examined using various structural correlation techniques like radial distribution function, angular distribution function, spatial distribution functions in order to rationalize the nonideal behavior. The correlated motion between ion pairs, quantified by ion pair lifetime time constants, was found to affect the self-diffusion coefficients and ionic conductivities for both the mixture systems.

471823 A Logic-Based Framework for Defining Force Field Usage Semantics and Atom-Typing Molecular Systems

Abstract:

The availability of forcefields for molecular simulation has reduced the effort researchers must devote to the difficult and costly task of determining the interactions between species, allowing them to instead focus on the motivating scientific questions. However, determining which parameters in a force field to use is still often a tedious and error prone task. This difficulty is related to the strong dependence of the parameters on the chemical context of the atoms, where the chemical context may depend on the local bonded environment of an atom in a molecule, the type of molecule(s) being considered, the phase of the molecule(s), etc. While tools for performing atom-typing exist, they are often tied to a specific force field or simulation software and typically employ a rigid hierarchy of encoded "if/else" statements where rules that identify more specialized atom types must be called in precise order, such that more general atom types are only chosen when more specialized matches do not exist. Rather than relying on rigid rule hierarchies that are difficult to maintain, debug, and evolve, we instead present a general, force field agnostic tool—Foyer—that relies upon logic-based annotations of chemical context. In Foyer, rigid rule hierarchies are replaced by an iterative process which allows rules to be more flexibly embedded within the actual force field files, alongside the parameter definitions, thus delivering annotations that are both human and machine readable. This also allow rules to be tested for logical consistency, independent of the chemical species they actually describe. Here, we present several case studies that demonstrate the forcefield annotation scheme and general software usage.

471441 Predicting Octanol-Water Partition Coefficient and Force Field Development for NDMA Via Molecular Simulations

Abstract:

N-Nitrosodimethylamine (NDMA) is industrial by-product or waste product of several industrial processes. In water treatment processes, chlorination of organic nitrogen containing wastewater can lead to the production of NDMA. NDMA's contamination of drinking water is of serious concern due to the difficulty in detecting it at low concentrations, and to the difficulty in removing it from drinking water. NDMA is highly toxic and is a suspected human carcinogen. Due to the hazard of carcinogens, a â€œhands offâ€ approach is necessary to study these substances and how they affect environment.Â The accurate estimation of the octanol-water partition coefficient will allow the prediction of the bioactivity of the contaminant in its immediate surroundings. The present study involves the force field development and predicting octanol-water partition coefficient for NDMA by molecular simulation methodology. We present Gibbs ensemble Monte Carlo simulations to obtain vapor liquid coexistence curves, Clausius-Clapeyron plots, critical properties and structural properties for NDMA. Molecular dynamics simulations are carried out to determine the octanol-water partition coefficient for NDMA via Bennett's Acceptance Ratio (BAR) method. At last, all the results obtained for NDMA with the optimized force field are compared to the available experimental data.

472047 Exploration of the Hierarchal Roles of Surface Features Facilitating the Adsorption of Biomolecules

Abstract:

Recent development in the anti-fouling properties of Self-Assembled Monolayers (SAMs) has largely focused on increasing the enthalpic association of a hydration layer along the interface of such surfaces with water. This has lead to the creation of surfaces that go beyond hydrogen-bonding with surrounding water molecules to ionic solvation, i.e. zwitterionic surfaces. However, an entropic penalty due to chain restriction also disfavors biomolecule-surface adsorption. While there isn't an outright dichotomy between the entropic point of view in SAM anti-fouling behavior and the more recent development of zwitterionic materials, an overarching theory on the protein adsorption resistance for surfaces is lacking. To isolate the effect of this entropic penalty amid changing packing densities, we performed all-atom (AA) molecular dynamics (MD) simulations of explicitly solvated systems of lysozyme and seven monomer length oligo (ethylene glycol) (OEG) SAMs. The rate of protein adsorption as well as the conformation and orientation of the protein upon adsorption were examined. It was found that chain spacing was a strong determinant in adsorption properties while chain mobility played a secondary role. In addition we performed AA MD simulations of a sodium styrene sulfonate (NaSS) surface in the presence of water and protein and investigated the features of the NaSS surface that selectively desorb extracellular matrix (ECM) proteins. This poster will discuss the aspects of the styrene sulfonate monomer that facilitated adsorption and end with a short perspective on how these features could be further optimized to selectively adsorb ECM proteins.

472135 Characterizing Protein Hydration to Inform Its Interactions with Ligands and Other Proteins

Abstract:

Protein interactions are crucial in numerous biological processes, including cell signaling, enzymatic function and disease pathology. Since all of biology happens in water, every bio-molecular binding process involves protein-water interactions being disrupted, and replaced by direct interactions between the binding partners. Thus, the key to predicting protein-protein interactions is to be able to accurately characterize the free energetics of protein-water interactions. Estimating protein-water interactions accurately and efficiently, however, has proved to be challenging, because proteins have incredibly complex surfaces that disrupt the inherent structure of water in countless different ways, which depend not only on the chemistry of the underlying protein surface, but also on the precise topography and chemical pattern of amino acids. Here, I will present our recent work on characterizing protein-water interactions using explicit water molecular simulations in conjunction with an unfavorable biasing potential that attempts to displaces water molecules away from the protein hydration shell. As the strength of the biasing potential is increased, protein-water interactions are systematically disrupted, resulting in the formation of cavities in the protein hydration shell. The order in which cavities appear in various regions of the protein contains a wealth of information: regions of the protein that interact weakly with water ought to dewet first, whereas those that are highly hydrophilic should hold on to their hydration waters even at large biasing potentials. Because the regions where cavities first appear have the weakest interactions with water, and correspondingly the highest surface energy, the protein is most likely to interact with other molecules through these regions. Furthermore, the size and the shape of the cavities that spontaneously appear in the protein hydration shell also inform the size and shape of an optimal hydrophobic ligand that will bind most strongly to the protein of interest. Our results could thus inform the design of customized ligands for protein separations using hydrophobic interaction chromatography (HIC).

472162 Understanding Wetting-Dewetting Transitions on Nanotextured Surfaces: Implications for Designing Surfaces with Robust Superhydrophobicity

Abstract:

Owing to a water contact angle greater than 150° and an effortless rolling off of water droplets, superhydrophobic surfaces exhibit self-cleaning, drag reducing and anti-icing properties. Such qualities of superhydrophobic surfaces make them revolutionary materials desired in myriad industrial applications and have spurred the discovery of numerous ways for texturing hydrophobic surfaces. It is now well established that superhydrophobicity derives from texturing of hydrophobic surfaces; the aversion of water for the hydrophobic surface prevents wetting of the surface texture, and stabilizes a Cassie state, in which a water droplet resides on the tips of the surface asperities. However, a key bottleneck in the widespread adoption of superhydrophobic surfaces has been the irreversible loss of superhydrophobicity, a consequence of wetting of the surface texture into a non-functional Wenzel state, under a variety of conditions ranging from elevated pressure and temperature to condensation of water in the texture. Furthermore, this wetting of the surface texture *i.e* the loss of superhydrophobicity is widely believed to be irreversible due to the large energy barriers that prevent the reverse (dewetting) transition. By using molecular dynamics simulations to characterize the free energetics and pathways of transition between the Cassie and the Wenzel states, our work strives to understand wetting-dewetting transitions on textured surfaces on a molecular level. Importantly, we illustrate how an understanding of dewetting pathways can inform the design of novel surface textures on which dewetting barriers vanish and superhydrophobicity can be spontaneously recovered. Such robust superhydrophobic surfaces, identified here for the first time, could find widespread application even under the most challenging conditions such as underwater operation or condensation heat transfer applications.

472423 Forming Auxetic Materials from Amorphous Networks

Abstract:

Auxetic materials, those with negative Poisson's ratios, are a promising candidate for impact control and mitigation. Recent work has shown that in amorphous solids, the contribution of any one bond or interaction to a given global property is not correlated with its contribution to other properties. We exploit this bond-level independence to independently modify the bulk and shear moduli of large-scale amorphous networks of nodes joined by bonds, creating auxetic materials. We develop straightforward computational models which classify the usefulness of a range of real materials towards creating amorphous auxetic materials. Given an appropriate choice of real material properties, simulated materials can be modified to be either fully incompressible or fully auxetic using a variety of boundary conditions. Such techniques can be applied to both two and three dimensional materials. Materials designed in simulation are created using 3D printing, and show excellent agreement with our models. Using both simulation and experiment, we probe the impact absorption properties of these materials.

472450 Atomistic Picture of the Growth Mechanism of Silicene Monolayers

Abstract:

Silicene, the Si based counterpart to graphene is predicted to demonstrate similar desirable features to graphene, possessing a Dirac cone in the band-structure and Fermi velocities comparable to graphene. Silicene has also been predicted to demonstrate several exotic properties such as the quantum spin Hall effect, chiral superconductivity and giant magnetoresistance. Furthermore, it has the benefit of being easy to integrate into current industrial device engineering processes. Silicene monolayers have been grown on Ag (111) and Ir (111) substrates and recently a working field-effect transistor has been demonstrated at room temperature. While it is possible to characterize the final monolayer structure, it is impossible to probe experimentally the evolving atomic structure during growth. We use large-scale molecular dynamics (MD) simulations to study the growth of silicene monolayers on Ag (111) and Ir (111) surfaces. Initially, individual Si atoms aggregate to form small clusters containing 3-6 Si atoms with different ring geometries. Over time, these rings merge to form larger clusters, but containing extended line defects and grain structures. Interestingly, structures grown on the Ag surface are prone to more defects, in particular extended regions of sub-surface growth, while the structures grown on Ir are relatively defect free. We also explore the effect of different growth conditions such as temperature and deposition flux on the structure of the monolayers. Finally, we study the mechanical and thermal properties of these *in-silico* grown monolayers and compare their performance to previously reported data for ideal structures. The processing-structure-property correlations we draw will enable experimental processes to be tailored to obtain target device properties.

471992 Sparse Sampling of Water Density Fluctuations

Abstract:

Water density fluctuations play an important role in understanding hydrophobic hydration and interactions. The rare low-density water density fluctuations at a solid-water surface have also been shown as a robust measure of surface hydrophobicity. However, characterizing the free energetics of such rare fluctuations requires computationally expensive, non-Boltzmann sampling methods like umbrella sampling, which limits the size of the volume in which the density fluctuations are characterized. Here we present a method that enables the sparse sampling of water density fluctuations, making it roughly 2 orders of magnitude more efficient than umbrella sampling. We circumvent the umbrella sampling requirement of overlap between adjacent biased distributions, by employing thermodynamic integration to estimate the free energy differences between biased ensembles. Leveraging the efficiency of the method, we are able to characterize water density fluctuations in the entire hydration shell of the protein, ubiquitin, a large volume containing an average of more than 600 waters, as well as enormous spherical volumes in bulk water containing up to 30,000 waters on average.

472503 Hybridization in DNA Self-Assembled Monolayer Studied with Atomistic Simulation and Free Energy Computation

Abstract:

The implantation of short nucleic acid fragments, ranging between 16-27 nucleic acids in length, on self-assemble surface has been shown to be a new class of novel diagnostic markers for diseases such as cancer. The sensitivity of such biochips is highly dependent on total hybridized amount of target strands from the bulk into the self-assembled monolayer (SAMs) of complementary strands grafted onto the substrate surface. Various factors affect the hybridization processes, such as surface morphology, surface coverage density, ion strength and target DNA concentration. To study the effects of those factors, we performed molecular dynamics simulations and free energy computation to study the morphology of DNA brush grafted onto a cristobalite substrate surface as well as its hybridization efficiency. Different DNA SAMs morphologies as a function of surface coverage density and the corresponding hybridized efficiency were examined. For the particular sequence we studied, an unhybridized DNA strand exhibits flexible coil, whereas a hybridized double strand displays a rigid rod structure. Counter ions in the system were attracted to the nucleic acid strands due to negative charged DNA chains. Moreover, lower diffusive activities of ions were observed inside condensed DNA SAMs. Hybridization free energy of DNA SAMs was predicted. Our research will facilitate the development of new biodevices based on nucleic acid hybridization.

472617 Development of Transferable Coarse-Grained Force Fields Via Multistate Iterative Boltzmann Inversion

Abstract:

Coarse-grained (CG) models are commonly used to access large system sizes and the long timescales required to observe interesting phenomena. While many methods and procedures exist for deriving CG force fields, the iterative Boltzmann inversion (IBI) method[1] has become a popular choice due to its straightforward nature and general applicability. However, the IBI method, like most approaches for deriving CG force fields, tends to generate force fields that demonstrate strong statepoint dependence. To overcome this limitation, we have developed a multistate extension to the iterative Boltzmann inversion CG potential derivation algorithm, termed multistate iterative Boltzmann inversion (MS IBI).[2] In MS IBI, a CG potential is iteratively updated to match target radial distribution functions across a range of thermodynamic and physical states, adding constraints to the force field optimization process. As such, the final potential demonstrates an increased level of transferability between states. The efficacy and advantages of using MS IBI to derive generally applicable CG force fields is further demonstrated in this work through two case studies. First, MS IBI is used to rederive the k-means CG water model[3] using both bulk and interfacial states, improving the quality of the forcefield.[4] Second, CG force fields for lipid systems of a ceramide sphingolipid and free fatty acid are derived and combined with surface wetting simulations.[5] The latter ensures hydrophobicity is accurately captured such that the contact angle of a water droplet on each surface matches detailed atomistic simulations, overall creating CG force fields that show close agreement atomistic and experimental results.

References

[1] 1. Reith, D., Putz, M. & MÃ¼ller-Plathe, F. Deriving Effective Mesoscale Potentials from Atomistic Simulations. J. Comput. Chem. 24, 1624â€"1636 (2003).

[2] Moore, T. C., Iacovella, C. R. & McCabe, C. Derivation of coarse-grained potentials via multistate iterative Boltzmann inversion. J. Chem. Phys. 140, (2014).

[3] Hadley, K. R. & McCabe, C. On the investigation of coarse-grained models for water: balancing computational efficiency and the retention of structural properties. J. Phys. Chem. B 114, 4590â€"9 (2010).

[4] Moore, T. C., Iacovella, C. R. & McCabe, C. Development of a coarse-grained water forcefield via multistate iterative Boltzmann inversion. Mol. Mod. and Sim.: Applications and Perspectives, in press (2016). arXiv:1509.07887

[5] Moore, T. C., Iacovella, C. R., Hartkamp, R. & McCabe, C. A Coarse-grained Model of Stratum Corneum Lipids: Free Fatty Acids And Ceramide NS. Submitted.

473305 Simulation-Based Design of Copolymer Sequence Using the Kinetic Monte Carlo Method

Abstract:

Free radical polymerization has a wide range of applications and continues to attract research interest as the demand for tailored specialty polymers grows. Sequence distribution, tacticity and composition of copolymers play a key role in understanding the reaction kinetics and the properties of the polymers [1]. One of the chief challenges in unraveling the relationship between sequence and properties is that experimental techniques allow only some aspects of sequence to be uncovered, but do not yet measure explicit sequence distributions. Extensive research effort is being put into modeling and simulation of free radical polymerization (FRP), in order to complement experimental efforts by revealing the full sequence information of all polymer chains. Among multiple modeling techniques, kinetic Monte Carlo (KMC) has been intensively exploited for simulating polymeric reactions, because of its capability of recording the explicit sequence of every polymer chain, enabling more detailed analysis of polymer properties [2, 3]. Nevertheless, compared to deterministic models, KMC is computationally much more expensive, limiting its applications for efficient design of sequence properties.

We developed a general KMC simulation framework that simulates free radical copolymerization, with the ability to keep track of the explicit sequence of polymer formed at any instant of the reaction, from which the statistics related to sequence distribution can be easily extracted, such as dyad or triad fractions and sequence length distribution [4]. The efficiency of this simulation model is maximized, reducing the simulation time while keeping accurate results for polymer sequence distributions and other properties including molecular weight distribution and copolymer composition [5]. The model was used for simulation based design for specific targets on sequence characteristics, and has the potential of being integrated into a simulation based optimization algorithm that searches for the required synthesis conditions for desired sequence properties.

[1] M. Al-Harthi, M. J. Khan, S. H. Abbasi, and J. B. P. Soares, "Gradient Copolymers by ATRP in Semibatch Reactors: Dynamic Monte Carlo Simulation," *Macromolecular Reaction Engineering,* vol. 3, pp. 148-159, 2009.

[2] Y. Mohammadi, M. Najafi, and V. Haddadiâ€Asl, "Comprehensive Study of Free Radical Copolymerization Using a Monte Carlo Simulation Method, 1," *Macromolecular theory and simulations,* vol. 14, pp. 325-336, 2005.

[3] P. H. M. Van Steenberge, D. R. Dâ€™hooge, M. F. Reyniers, and G. B. Marin, "Improved Kinetic Monte Carlo Simulation of Chemical Composition-Chain Length Distributions in Polymerization Processes," *Chemical Engineering Science,* vol. 110, pp. 185-199, 2014.

[4] V. R. Regatte, H. Gao, I. A. Konstantinov, S. G. Arturo, and L. J. Broadbelt, "Design of Copolymers Based on Sequence Distribution for a Targeted Molecular Weight and Conversion," *Macromolecular Theory and Simulations,* vol. 23, pp. 564-574, 2014.

[5] H. Gao, L. H. Oakley, I. A. Konstantinov, S. G. Arturo, and L. J. Broadbelt, "Acceleration of Kinetic Monte Carlo Method for the Simulation of Free Radical Copolymerization through Scaling," *Industrial & Engineering Chemistry Research,* vol. 54, pp. 11975-11985, 2015.

477521 Coarse Grained Model of Conformational Disorder Effects on the Electronic States of Poly(3-hexylthiophene) Chains

Abstract:

Semiconducting polymers are soft materials with many conformational degrees of freedom. Limited understanding of how conformational disorder affects their optical and electronic properties is a key source of difficulties limiting their widespread usage in electronic devices. We develop a coarse-grained approach based on the tight binding approximation to model the electronic degrees of freedom of polythiophene chains, taking into account atomic coordinates changes. Particularly important are dihedral rotations and moiety distortions, which are known to disrupt extended electronic states. The initial stage of our model successfully captures the effects of dihedral rotations on the valence and conduction states of the chains. We extend the model to include the effects of thiophene ring vibrational distortions. We use density functional theory (DFT) to characterize the coupling between the ring normal modes and charge carriers. Our model combined with the variational method is used to predict the formation and the spatial extent of ring-stabilized polarons in poly(3-hexylthiophene). As a way of assessing the relative importance of these polaronic effects to transport, we compute the induced change in carrier effective mass, and compare the polaron size to the carrier localization length resulting from thermal dihedral fluctuations.

458118 Antimicrobial Mechanism of Action of Class IIb Bacteriocins Investigated through Molecular Simulations

Abstract:

The rise of antibiotic resistance over the last decade is alarming. Studies emerging worldwide warn that if no significant measures are taken, we will be soon entering a post-antibiotic era where even minor infections could prove deadly. In response, there has been an extensive effort to identify and develop new antibacterial agents. Bacteriocins are antimicrobial peptides (AMPs) produced by bacteria as a part of their defense mechanism against antagonistic organisms and can be used as an alternative to conventional antibiotics. In order to exploit their full potential we need first to understand their mechanism of action.

In the case of class IIb bacteriocins, two different peptides are required in order to exhibit bactericidal activity. It is hypothesized that, in order to function, these two peptides must form a dimer. It has been shown that class IIb bacteriocins disrupt the membrane of the target cells. However, it is unknown if they act on their own, or if they act through a receptor-mediated mechanism as has been suggested for other class II bacteriocins.

We use plantaricin EF (plnEF) as a representative class IIb bacteriocin. PlnEF is a heterodimeric bacteriocin consisting of two peptides: plantaricin E (plnE) and plantaricin F (plnF). We performed atomistic molecular dynamics (MD) simulations of these peptides in model membranes. In combination with experimental studies, we attempt to shed light on their mechanism of action. We initially studied the behavior of the peptides on the surface of the membrane [1]. We identified the residues that facilitate peptide-peptide and peptide-membrane interactions. We also simulated various mutants of the peptides. We compared the results with the experimentally observed activity of the mutants, which allowed us to propose structure-activity relationships.

Guided by the above-detailed findings and new experimental results, we designed a possible transmembrane dimer conformation. Our simulations revealed that the peptides interact through a GxxxG motif, which is also present in most other class IIb bacteriocins. Moreover, persistently stable hydrogen bonds and cation-Ï€ interactions strengthened the association between the peptides and stabilized the dimer inside the membrane. The simulated behavior of the peptides agreed well with experimental findings and provided a better interpretation on an extensive mutational analysis [2]. After long atomistic MD simulations, the dimer was stable and resembled the behavior of other well-characterized transmembrane helical domains that contain the GxxxG dimerization motif. Advanced sampling techniques were employed in order to explore the dimerization free energy landscape and to investigate whether these class IIb bacteriocins have the ability to induce ion transport by themselves.

We believe that our studies can shed light on the antimicrobial mechanism of action of class IIb bacteriocins and provide valuable information for the design and optimization of new antibiotic drugs.

References:

[1] P.K. Kyriakou, B. Ekblad, P.E. Kristiansen, Y.N. Kaznessis, Interactions of a class IIb bacteriocin with a model lipid bilayer, investigated through molecular dynamics simulations., Biochim Biophys Acta. 1858 (2016) 824â€"35.

[2] B. Ekblad, P.K. Kyriakou, P.E. Kristiansen, C. Oppergard, Y.N. Kaznessis, J. Nissen-Meyer, Structure-Function Analysis of the Two-Peptide Bacteriocin Plantaricin EF., (under preparation).

461255 Carbohydrate Recognition Mechanisms in Cellulose-Specific Type B Carbohydrate Binding Modules

Abstract:

Effective enzymatic degradation of crystalline polysaccharides requires a synergistic cocktail of hydrolytic enzymes tailored to the wide-ranging degrees of substrate crystallinity. To accomplish this type of targeted carbohydrate recognition, nature produces multi-modular enzymes, having at least one catalytic domain appended to one or more carbohydrate binding modules (CBM). Type B CBMs encompass several families (i.e., protein folds) of CBMs that are generally thought to selectively bind oligomeric polysaccharides; however, a subset of cellulose-specific CBM families (4, 17, and 28) appear to bind non-crystalline cellulose more tightly than oligomers and in a manner that discriminates between surface topology. In this study, we investigated the molecular-level origins of oligomeric and non-crystalline carbohydrate recognition in cellulose-specific Type B CBMs using molecular dynamics simulation and free energy calculations. Examining two CBMs from each of the three families, we describe how protein-ligand dynamics contribute to observed variations in oligomeric binding affinity within the same CBM family. Comparisons across the three CBM families identified factors leading to modified functionality prohibiting competitive binding, despite sequential and specificity similarities. Using Free Energy Perturbation with Hamiltonian Replica Exchange Molecular Dynamics, we examined the hypothesis that the open topology of the binding grooves in families 17 and 28 necessitate tight binding of an oligomer, while the more confined family 4 binding groove does not require the same degree of tight binding. Finally, we elucidated the mechanisms of non-crystalline carbohydrate recognition by modeling a family 28 CBM complexed with a partially descrystallized cellulose substrate. Molecular simulation provided structural and dynamic data for direct comparison to oligomeric modes of carbohydrate recognition, and thermodynamic integration was used to determine ligand binding free energy. Comparing both protein-carbohydrate interactions and ligand binding free energies, which were within error of experimental values, we validated the correlation of high- and low-affinity binding sites with non-crystalline and oligomeric binding, respectively. Our study provides an unprecedented level of insight into the complex solid and soluble carbohydrate substrate recognition mechanisms of CBMs, the findings of which hold considerable promise for enhancing lignocellulosic biomass conversion technology.

461829 Gold Nanoparticles Change Human Serum Albumin without Denaturing It: A Computer Simulation Study

Abstract:

Excitement about the applications envisioned for nanomaterials in energy, medicine and electronics is tempered by concerns about their toxicity. The toxicity of nanomaterials is believed to be closely related to their ability to denature proteins. This work investigates how binding to a gold nanoparticle with a diameter of 4.0 nm affects the structure and flexibility of human serum albumin. Nanomaterials are considered less toxic if they let proteins remain folded. However, the results of our computer simulations argue that nanomaterials can still induce toxicity even if they do not denature proteins. We identify four binding complexes between the gold nanoparticle and human serum albumin in a brute-force search process using an implicit-solvent coarse-grained model. We then investigate the structural and dynamic properties of the albumin protein within the complexes in explicit solvent using atomistic molecular dynamics simulations. The results show that although the albumin protein remains folded in the simulations, many of its residues change their flexibility and secondary structure. Analysis of these changes shows that the binding of a nanoparticle can alter the flexibility and secondary structure of a protein significantly without denaturing it. The allosteric effect plays a vital role in spreading the impact of a nanoparticle throughout the whole protein. Our simulations suggest that keeping a protein folded does not guarantee a nanoparticleâ€™s safety.

466492 Enhanced Sampling of Protein Structures at the Air-Water Interface

Abstract:

Proteins at the air-water interface are of importance for several relevant biological and industrial purposes including the formation of biofilms and the production of foods containing large air bubbles. The unique way in which water orders itself at interfaces with air and the nature of air as a disordered hydrophobic medium cause proteins to exhibit large structural differences between solution and interfacial structures. One of the few experimental methods that provides information about the structure and orientation of proteins at the air-water interface is sum frequency generation spectroscopy (SFG). One shortcoming of SFG is that diverse structures can produce similar spectra. To overcome this challenge, molecular simulation can be utilized to provide an ensemble of structures that are likely present at the interface. In this study, we use well-tempered metadynamics to uncover the most probable structures of four model peptides at the air-water interface. Then, the spectra of these structures are calculated and compared to experiments. Several different protein and water force field combinations are examined to determine which model is most consistent with experiments. We find that CHARMM-family force fields prefer to remain near the solution structure while AMBER-family force fields exhibit much more conformational diversity. This information and the methods we develop can be used to guide future simulations of proteins at the air-water interface and to provide detailed structures that accurately represent SFG data.

466621 Binding Affinities of Amino Acids on Graphene: Assessment of Force Fields

Abstract:

Graphene–protein interactions are central to understanding the biocompatibility of graphene (GRA). Understanding these interactions is important for many diverse applications such as biosensors, theranostics, cosmetics and electronics. The mechanisms and driving forces that govern protein and peptide interactions with GRA are not completely understood. Recent experimental and simulation studies indicate a dominance of aromatic amino acids and arginine in mediating the GRA–peptide interactions. However, in our previous combined experimental and simulation study of albumin adsorption on GRA, we observed no such dominant effect from any of the amino acids.[1]

Furthermore, recently reported binding strengths for these amino acids on GRA have been inconsistent. In general, most of the force fields employed are developed to represent proteins in aqueous media but not at material interfaces, such as graphene and self-assembled monolayers. In a recent study, the free energy of adsorption of peptides on silica surfaces using CHARMM force field was observed to deviate as large as 5-9 kcal/mol with the experimental data. This highlights the need to carefully evaluate the different force-fields available to describe protein systems in the context of GRA-protein interactions. To this end, we have performed extensive simulations to calculate the free energy of adsorption of amino acids on GRA. Specifically, we have used umbrella sampling method to determine the potential of mean force of all twenty amino acids using three of the widely employed force fields—Amberff99SB-ILDN, CHARMM-36 and OPLS-AA/M in explicit water. We compare the calculated free energy of adsorption with the available experimental data. In our talk, we will present these detailed results discussing the trends and their implications in understanding GRA–peptide interactions.

---

[1]

Sengupta, Bishwambhar, Wren E. Gregory, Jingyi Zhu, Siva Dasetty, Mehmet Karakaya, Jared M. Brown, Apparao M. Rao, John K. Barrows, Sapna Sarupria, and Ramakrishna Podila. "Influence of carbon nanomaterial defects on the formation of protein corona." *RSC Advances* 5, no. 100 (2015): 82395-82402.

462620 How CO_2 Diffuses into the Active Site of Carbonic Anhydrase: A Study By Molecular Dynamics Simulation and Multi-Scale Markov-State Modeling

Abstract:

While a significantly enhanced capture of CO_2 has been accomplished with the aid of carbonic anhydrase (CA), which catalyzes the conversion of CO_2 to bicarbonate at an extremely fast rate, the molecular mechanism for the diffusion of CO_2 from the bulk aqueous phase into the active site of CA is not yet established. Here, we present theoretical results from molecular dynamics (MD) simulation and multi-scale Markov-state modeling (MSM) of the CO_2 diffusion process using carbonic anhydrase (Î±-human CA II, EC 4.2.1.1) as a model enzyme. CO_2 diffusion along the natural gateway of CA was illustrated with a two-dimensional free energy landscape along two reaction coordinates, i.e., the CO_2 distance (center of mass) from a water molecule attached to the zinc complex, and the orientation angle formed by O_{H2O}-C_{CO2}-O_{CO2} atoms. Two distinct diffusion modes were identified using the Perron cluster cluster analysis (PCCA) in combination with the flux analysis for the transition rate matrix of the coarse-grained Markov state model (CG-MSM). The simulation results for K_m, one key parameter of the Michaelis-Menten equation, agreed well with that from the experimental data. Moreover, the multi-scale MSM indicated a preferential accumulation of CO_2 at the hydrophobic pockets inside CA. A good understanding of the molecular events underpinning enzyme kinetics will be helpful for both molecular re-engineering of CA by either chemical or genetic methods as well as for process intensification or renovation for CA-assisted CO_2 capture and utilization.

468048 Conformational Ensemble of Islet Amyloid Polypeptide in a Membrane Environment

Abstract:

Islet Amyloid Polypeptide (IAPP) is a 37 residue intrinsically disordered protein implicated in the pathology of Type II Diabetes. Intrinsically disordered proteins have no definite structure, but rather sample an ensemble of structures. IAPP, also called Amylin, has been found to form amyloid fibrils due to misfolding and aggregation which is accelerated in the presence of lipid bilayers. When Amylin is in a membrane environment, its conformational ensemble is altered in ways which promote the formation of these amyloid fibrils. Using atomistic molecular dynamics simulations, we probe the conformational ensemble of Amylin in bulk, as well as in a membrane environment to observe the effects which cause the increased amyloidogenic activity.

459214 Investigation on the Pathway of Disulfide Bonds Reduction on Human Defensins Using Molecular Dynamics Simulations

Abstract:

Defensins are cationic cysteine-rich small molecules with molecular masses ranging from 3 to 5 kDa. They are a critical part of the innate immune system that provides an initial antimicrobial barrier for mucosal surfaces such as the surface of the eyes, airways, lungs, and skin. They are known as 'natural antibiotics' as they have broad spectrum antimicrobial activity against both Gram-positive and Gram-negative bacteria, and even against bacteria that have demonstrated resistance to the currently used antibiotics. Mammalian defensins are classified into a, b, q categories based on their size and disulfide bonding pattern. Human a defensin type 5 (HD5) is mainly produced by Paneth cells in the crypts of the small intestine. The peptide has 6 cysteine residues forming 3 disulfide bridges: Cys3-Cys31, Cys5-Cys20, Cys10-Cys30. Human b defensins (HBD) are produced by epithelial cells, for example in gut and lung. HBD type 3 (HBD-3) is a newly discovered defensin, which also has 6 cysteine residues, and forms three disulfide bridges: Cys11-Cys40, Cys18-Cys33, and Cys23-Cys41.

The disulfide bonds of defensins break at reduced condition, and human defensins show diverse activity dependence on their disulfide bridge structure. For example, for human b defensin type 3 (HBD-3), its disulfide bridge status doesn't seem to affect its anti-bacterial activity, but does have influences on its chemotactic activity; while for HD-5, reduction of all three disulfide bridges may contribute to its wide variety of antimicrobial, antivirial and immune modulating activities. In order to understand the different effects of disulfide bridges on functional activities of different defensins, it is important to find out the pathway of disulfide bonds reduction at reduced condition.

Since there is only experimental data available for HD5 disulfide bridge reduction pathway, a simulation strategy including performing all atom molecular dynamics simulations using both CHARMM and NAMD programs was carried out on HD5 first. The disulfide bond reduction pathway on HD5 was analyzed, which predicted results consistent with available experimental data. Using the same simulation strategy, the disulfide bond breaking order on HBD-3 was predicted. It was found out that the majority of HD5 unfolds by initial reduction of C5-C20, followed by C10-30 and C3-C31, while under the reduced condition, the HBD-3 disulfide bond on C18-C33 would always break first. Those results can shed light on understanding the different functional activity dependence on the disulfide bridge status for HD5 and HBD-3.

472079 Predicting Protein-Protein Interaction Interfaces, Hot Spots, and Binding Affinities

Abstract:

Protein interactions play an important role in various biological processes, such as cell signaling, drug delivery and disease. Predicting them requires an understanding of protein-water interactions because every protein binding process involves the protein-water interactions being disrupted, and replaced by direct interactions between the binding proteins. Characterizing protein-water interactions accurately, however, has proved to be challenging, because these interactions depend on not only the nanoscale topography, but also the precise chemical pattern presented by the protein surface. Here we propose a novel method for characterizing the strength of protein-water interactions and predicting the interface through which proteins interact with one another. Our approach relies on the application of an unfavorable biasing potential to the entire protein hydration shell. The response of the hydration shell waters to the applied potential contains a wealth of information, which we exploit. In particular, we identify the regions from which the protein hydration waters are displaced, resulting in the formation of cavities. The regions where cavities appear most readily ought to be the most hydrophobic, and we find that such thus regions are the most probable interfaces through which proteins bind to other proteins or non-polar ligands. Our method thus enables us to determine protein interaction interfaces, which we then use to determine binding free energies. Finally, our approach also enables determination of protein residues that contribute the most to the binding affinity, the so-called hot-spots.

475359 Best Practices in Collaborative Software Development: Lessons Learned in the Rosetta Commons

Abstract:

The Rosetta software (http://rosettacommons.org) enables structure prediction and design of biomolecules, including proteins, nucleic acids, molecular complexes, and symmetric assemblies. It is developed jointly by a collaboration of 50 research groups distributed around the world. We share the computer code on GitHub, where there are hundreds of active developers. New methods are shared instantaneously among developers, accelerating the pace of scientific discovery. To maintain the effectiveness and collegiality of this collaboration, we have built an extensive set of tests, code development tools, training courses, and documentation, and we use social media, team scrums, lab visits, an annual conference and organized outreach. In this talk, I will present our scientific environment, intellectual property agreement, test systems, communication and training tools. Finally, I will summarize what we have learned about best practices for building communities of scientific software developers.

469131 Validation of Computational Models and Codes

Abstract:

Validation and verification are companion processes in assuring the quality and reproducibility of numerical simulations. While verification tends to focus on software, validation is intended to ensure that the model we are simulating accurately reflects the behaviors of the system we are studying. Failure to validate numerical modeling can have catastrophic consequences, as exhibited by the 1998 Mars Climate Orbiter mission, as well as other engineering failures. We discuss the basic principles underlying the validation process, and present examples from our research groups on how the validation process can be applied to molecular simulations.

469959 *Pysimm*: A Python Package for Simulation of Amorphous Polymeric System

Abstract:

There is a great need for new easy-to-use applications to perform computational studies of polymeric materials. Whether the goal is to educate new researchers or expand accessibility to purely experimental researchers, one prohibitive hurdle is often the time spent learning the intricacies associated with complex simulation packages or high performance computing techniques. *pysimm*, short for python simulation interface for molecular modeling, is a python package designed to facilitate structure generation and simulation of molecular systems. *pysimm* provides a collection of different simulation tools while offering smooth integration with highly optimized third party software for specialized tasks. The platform provided by *pysimm* has helped the rapid development of new molecular modeling applications specifically in the area of amorphous polymeric systems, an area that can benefit greatly from more simulation tools.

The modular tools in the *pysimm* package provide a toolbox for researchers to build applications with complex workflows through easy to use functions and fully object oriented representations of molecular systems. *pysimm* interfaces with existing software, such as the LAMMPS simulation package, to perform expensive or highly specific computations. The core structure of *pysimm* is described here as well as examples of how *pysimm* can be easily used to build applications.

In general, polymeric material properties depend greatly on the specifics of each polymeric sample such as the molecular weight distribution, degree of branching, network structure, tacticity and monomeric composition. A given application that excels in preparing polymer structures with one given morphology may not be efficient creating another one. An example showing the development of a new computational polymer growth algorithm using *pysimm* designed to control molecular weight and build copolymers is highlighted here.

An additional area of interest is using *pysimm* to create cloud-based web applications. Developing applications for the cloud has the benefit of allowing researchers to control the version of their software use as well as limit the complications users might have because of hardware or software incompatibilities. The union of *pysimm* with freely accessible cloud-based web applications will bolster the growth of open-access databases for predictive amorphous polymer simulations with robust and modular tools and validation with experimental results.

458831 Automated Event Detection and Rate Constants for Complex Systems with Reaxff

Abstract:

Complex chemical engineering processes are ultimately determined by molecular processes, often in complex systems. Reactive molecular dynamics techniques such as ReaxFF afford insight at the atomistic level in complex chemical processes such as combustion, heterogeneous catalysis, battery discharge and material degredation.

We will discuss how automated detection tools give both qualitative and quantitative insight in chemical reactions and physical events such as adsorption and diffusion. Chemical reaction pathways for complicated reactions can be mapped out and browsed and reaction rates automatically determined.

Accelerated dynamics techniques give access to even longer time scales at the molecular level, enabling better integration with mesoscale tools for multi-scale methods used for modeling chemical engineering problems. We will introduce a few available acceleration techniques, in particular the force bias Monte Carlo.

Finally, we will highlight the advantage of having DFT, DFTB, ReaxFF and continuum methods for fluid thermodynamics (COSMO-SAC + COSMO-RS) in a comprehensive software package, the ADF Modeling Suite. A single integrated graphical interface and extenisve cross-module scripting tools ensure ease of use across all modeling tools.

460452 Freud: A Software Suite for High-Throughput Simulation Analysis

Abstract:

Computer simulation is an indispensable tool for the study of a variety of systems, including biological and nanoscale systems. As simulations scale to increasingly powerful supercomputing clusters, so too does the size of the data produced, as well as the difficulty in analyzing these data. We present *Freud*, a simulation analysis software suite for efficient analysis of simulation data. Freud makes no assumptions about the system being analyzed, allowing for general analysis methods to be applied to nearly any type of simulation. Written with modern simulation and data analysis workflows in mind, Freud provides a Python interface to fast, parallel C++ analysis routines, and runs efficiently on laptops, workstations, and supercomputing clusters. Freud includes standard analysis methods such as the radial distribution function, as well as new methods including the potential of mean force and torque and local crystal environment analysis. Used in conjunction with simulation software such as HOOMD-Blue, Freud can analyze simulations in real time, allowing for smart simulations that adapt to the current state of the system, enabling the study of phenomena such as nucleation and growth.

450593 PB-[S]AM, a Novel Solution to the Poisson-Boltzmann Equation for Applications Ranging from Protein Simulations to Polymer Membrane Design

Abstract:

While molecular dynamics (MD) simulations routinely and beneficially address questions concerning detailed structure and kinetic rates of assembly over fast timescales, there is another set of problems on the mesoscale that surpass the limits of timescale and size for MD. Fortunately, coarse-graining (CG) the participating nanoscale structures and their environment, combined with simulations using stochastic dynamics, can be just as insightful as the all-atom dynamics.

Electrostatic forces often dominate interactions in applications ranging from recognition events for biomolecular complex formation to proton transport in membranes. For this reason, a realistic simulation must include an accurate treatment of bulk electrolytes for large-scale systems. This can be achieved using continuum mean-field theory with the Poisson-Boltzmann (PB) equation, which, under the low-field condition, may be simplified to the linearized Poisson-Boltzmann equation (LPBE).

Our group has achieved a fundamental result in deriving the first completely general analytical solution to the LPBE for computing screened (salty) electrostatic interactions between arbitrary numbers of nanoscale spheres of arbitrarily complex charge distributions, separated by arbitrary distance. This analytical solution, PB-AM, serves as the foundation of a semi-analytical approach to solving the LPBE for any shape (PB-SAM). We have recently incorporated PB-SAM into a Brownian Dynamics approach to create a robust CG simulation tool to study a range of biological systems.

Overall, PB-SAM has many benefits when compared to other PB solvers including accuracy, flexibility and memory management. We are currently working to integrate this package into the distributed software package APBS, and are continually seeking methods to improve the software. Additionally, we are continually working to improve PB-SAM's efficiency, leveraging additional parallelization through further OpenMP implementations and exploring the option of MPI parallelization. We are also introducing the 3- or 4-body approximation to speed-up the n-body polarization.

This program has been used to examine protein-protein interactions in a crowded environment, and will be used in the future to explore fundamental issues in performance properties of polymer electrolyte membranes, particularly in regards to water flooding and insufficient ion transport, which have prevented the widespread commercialization of it as a replacement for the common battery.

462953 Multi-GPU Electrostatics and Rigid Bodies in HOOMD-Blue 2.0

Abstract:

We report on the addition of a scalable particle-particle particle-mesh (PPPM) implementation to HOOMD-blue 2.0, which is capable of calculating the long-ranged Coulomb interaction efficiently on several GPUs using spatial domain decomposition and the Ewald summation. The method is based upon a scalable implementation of the Fast Fourier Transform (FFT) in MPI and on the CUDA FFT library (CUFFT) and uses internal block-cyclic data representations and vector radix transforms to speed up the calculation. We also present on a redesigned capability of HOOMD-blue to integrate the equations of motion for rigid bodies composed of smaller particles, which readily scales to many GPUs and allows for easy creation of copies of identical rigid bodies.

464854 Scientific Data Management with Signac

Abstract:

Researchers in the field of computational physics, chemistry, and material science are regularly posed with the challenge of managing large and heterogeneous data spaces. The amount of data increases in lockstep with computational efficiency multiplied by the amount of available computational resources, which shifts the bottleneck within the scientific process from data acquisition to data post-processing and analysis. We present a framework designed to aid in the integration of various specialized formats, tools and workflows. The signac framework provides all basic components required to create a well-defined and thus collectively accessible data space, simplifying data access and modification through a homogeneous data interface. The framework's data model is designed not to require absolute commitment to the presented implementation, simplifying adaption into existing data sets and workflows. This approach not only increases the efficiency of the production of scientific results, but also significantly lowers barriers for collaborations requiring shared data access.

466558 Automated Molecular Modeling Via Optimization of Force Field Parameters

Abstract:

One goal of simulations is to predict chemical properties such that experiments can be minimized. A critical part of this is the reliable generation of molecular models. Consequently, we have developed several modular program packages that allow to parameterize entire force fields semi- and fully automatically. Various interlinked modeling resolutions are addressed. For intramolecular interactions, a force-field optimization package named Wolf2Pack [1] has been developed that transfers knowledge gained from quantum mechanics to Newtonian-based molecular models. For intermolecular interactions, especially Lennard–Jones parameters, a modular optimization toolkit has been developed that combines global and local optimization algorithms. Global coarse optimization is performed by the meta-model based tool CoSMoS [2], while local finer optimization is done by the gradient-based optimization workflow named GROW [3] or by a derivative-free method called SpaGrOW. The overall goal of our toolkits is to realize an easy, efficient, systematic and user-friendly development of reliable force-field parameters in reasonable time. The various tools are needed and interlinked since different stages of the optimization process demand different courses of action. In this presentation, we will present an overview of these toolkits, our approach towards parameterization (e.g. target observables, transferability, data archiving), and discuss some interesting issues that we have encountered along the way.

[1] Krämer-Fuhrmann, Ottmar; Neisius, Jens; Gehlen, Niklas; Reith, Dirk; Kirschner, Karl N. (2013): Wolf2Pack--portal based atomistic force-field development. In: *J Chem Inf Model* 53 (4), S. 802–808. DOI: 10.1021/ci300290g.

[2] Krämer, Andreas; Hülsmann, Marco; Köddermann, Thorsten; Reith, Dirk (2014): Automated parameterization of intermolecular pair potentials using global optimization techniques. In: *Computer Physics Communications* 185 (12), S. 3228–3239. DOI: 10.1016/j.cpc.2014.08.022.

[3] Hülsmann, Marco; Köddermann, Thorsten; Vrabec, Jadran; Reith, Dirk (2010): GROW: A Gradient-based Optimization Workflow for the Automated Development of Molecular Models. In: *Computer Physics Communications* 181, S. 499–513. DOI: 10.1021/ci300290g.

457895 Efficient Neighbor List Calculation for Molecular Simulation of Colloidal Systems Using Graphics Processing Units

Abstract:

Soft matter systems are characterized by a significant disparity in relevant length and time scales between constituent components. For example, colloidal particles (nanometers to micrometers in diameter) in solution are separated in size by several orders of magnitude from an atomistic description of the solvent. Molecular dynamics (MD) simulations retaining full atomistic detail of the solvent can become intractable because many solvent atoms must be included to model only a few colloidal particles. Moreover, the time scales associated with the degrees of freedom of the solvent are generally much shorter than the relatively slow motion of the larger colloids. This means that these simulations require very short MD time steps to faithfully capture the dynamics of the solvent, and many such steps are required to observe any appreciable dynamics of the colloids.

Calculation of the forces between particles, and in particular, nonbonded pair interactions, dominates the computational time of the MD algorithm. In order to accelerate these calculations, pair potentials are frequently truncated at a finite cutoff distance, and a neighbor (Verlet) list of particles within each othersâ€™s cutoffs is maintained to reduce the number of distance evaluations between particles that are performed. The calculation of the neighbor list itself is accelerated using a structure such as a spatially-binned cell list that reduces the overall computational complexity of the algorithm.

We present an algorithm based on linear bounding volume hierarchies (LBVHs) for computing neighbor (Verlet) lists using graphics processing units (GPUs) for colloidal systems characterized by large size disparities. We compare this in the HOOMD-blue simulation package to a GPU implementation of the current state-of-the-art CPU algorithm based on stenciled cell lists. We report benchmarks for both neighbor list algorithms in a Lennard-Jones binary mixture with synthetic interaction range disparity and a realistic colloid solution. LBVHs outperform the stenciled cell lists for systems with moderate or large size disparity and dilute or semidilute fractions of large particles, conditions typical of colloidal systems. The LBVH neighbor list algorithm has recently been applied to study transport and structure formation in colloidal dispersions under flow.

472818 Methods for Combining Experimental Data and Molecular Simulations into Hybrid Models

Abstract:

Creating models that are consistent with experimental data is an essential task in computational modeling. This is generally done by iteratively tuning the input parameters of a simulation to match experimental data. An alternative method is to bias a simulation to match experimental data leading to a hybrid model composed of the original model and biasing terms derived from experimental data. Recently, we've developed two methods that can be mathematically shown to bias a simulation as little as possible to match experimental data. One method is for matching PMFs in collective variables space (e.g., RDF) and another for ensemble averages. I'll report progress on biasing dynamical quantities as well and present results on applying these methods in both ab initio models and molecular dynamics simulations.

455267 Using Molecular Dynamics to Understand Transmission Electron Microscopy

Abstract:

Modern transmission electron microscopy (TEM) enables the imaging of many complex systems (cells, macromolecules, proteins, or nanoparticles) in near-native environments. In order to improve interpretation of TEM images it is necessary to predict features and ideal imaging conditions ahead of time. To this end virtual electron microscopy techniques are needed. We present a combination of molecular dynamic simulations with image simulations and apply this technique to a variety of nanoparticle systems. We initially present PbS nanoparticles under different imaging conditions to validate the methodology. General insight is obtained on the influence of different experimental conditions for image quality.

Furthermore we apply the technique to model Li Ion battery in operation conditions as well as to understand the role of solvation forces on the preferential attachment of nanoparticles in liquid systems.

1. David A. Welch, Roland Faller, James E. Evans, Nigel D. Browning: Simulating Realistic Imaging Conditions For In-Situ Liquid Microscopy Ultramicroscopy 135 36-42 (2013)

2. B. Layla Mehdi, E. Nasybulin, J. Qian, C. Park, David A. Welch, Roland Faller, H. Mehta, W. A. Henderson, W. Xu, C. M.Wang, James E. Evans, J. Liu, J. -G. Zhang, K. T. Mueller, and Nigel D. Browning: "Observation and Quantification of Nanoscale Processes in Lithium Batteries by Operando Electrochemical (S)TEM" Nano Letters 15(3) 2168-2173 (2015)

3. David A. Welch, B. Layla Mehdi, Hannah J. Hatchell, Roland Faller, James E. Evans, Nigel D. Browning: "Using Molecular Dynamics to quantify the electrical double layer and examine the potential for its direct observation in the in-situ TEM" Advanced Structural and Chemical Imaging 1 1 (2015)

4. David A. Welch, Taylor Woehl, Chiwoo Park, Roland Faller, James E. Evans, Nigel D. Browning "Understanding the Role of Solvation Forces on the Preferential Attachment of Nanoparticles in Liquid" ACS Nano 10(1) 181-187 (2016)

5. Ali Esmaieeli Sikaroudi, David A. Welch, Taylor Woehl, Roland Faller, James E. Evans, Nigel D. Browning, Chiwoo Park "Directional Statistics of Preferential Orientations of Two Shapes in Their Aggregate and Its Application to Study Preferential Attachment of Nanoparticles" (submitted)

470473 Pushing the Boundaries of Multiscale Simulation to Model Proton Transport in SERCA

Abstract:

The sarcoplasmic reticulum Ca^{2+}-ATPase (SERCA) calcium pump is ubiquitous in cells, exchanging Ca^{2+} ions for protons (powered by ATP) to maintain homeostasis and regulate muscle contraction. Its dysfunction has been indicated in multiple diseases including heart failure, respiratory diseases (e.g. cystic fibrosis), diabetes, and Alzheimer's disease. While SERCA structures have been solved and multiple experimental and computational studies have elucidated calcium transport mechanisms, a proton transport mechanism has yet to be determined. Computational molecular modeling could allow investigation of such mechanisms, but to do so required advances in multiscale modeling. The full system requires including the ~1000 residues of SERCA, any associated ligands (such as its natural inhibitor phospholamban), a lipid bilayer, and surrounding water, with sub-Angstrom resolution in the regions involved in proton transport. The dynamics of the conformational flexibility of the protein, the majority of which resides outside the lipid bilayer, are orders of magnitude slower than the fast but infrequent dynamics of proton transport. To span the time and length scales required to sample the ensemble of relevant protein conformations with the resolution needed to simulate proton transport, we combined enhanced sampling methods with our multi-scale reactive molecular dynamics (MS-RMD), which allows modeling of the bond-formation and cleavage required for proton transport at significantly lower computational cost than *ab initio* molecular dynamics or QM/MM simulations. Advances in applying these methodologies to bio-molecular systems have allowed us to determine a mechanism for passive proton permeability through SERCA in its inactive, phospholamban-bound state, which had been hypothesized to explain how the sarco- or endoplasmic reticulum maintains equilibrium with the cytosol during calcium release via separate calcium channels. The methods described here continue to expand the power of simulation to model more complex systems of interest.

461300 Modeling the Nucleation of Crystals of Ionic Liquids in the Bulk and Near Graphitic Surfaces

Abstract:

Solidification of ionic liquids (ILs) is used in the synthesis of optically-active and magnetic nanomaterials based on ILs. However, very little is understood about the mechanism of nucleation of solid phases of ILs, mainly because most of their applications involve liquid phases. Here we studied the nucleation of crystals of two common ILs, 1,3-dimethylimidazolium chloride [dmim$^+$][Cl$^-$], and 1-butyl-3-methylimidazolium chloride [bmim$^+$][Cl$^-$], from their supercooled liquid phases in the bulk and in contact with graphitic surfaces. The string method in collective variables[1,2] was used in combination with Markovian milestoning with Voronoi tessellations[2,3] and order parameters for molecular crystals[4] to sketch a minimum free energy path connecting the supercooled liquid and the crystal phases, and to determine the rates involved in the nucleation process. This particular combination of methodologies is well suited to study nucleation of ILs, as these systems exhibit very slow dynamics and very large simulation box sizes are required in order to avoid finite-size effects. We will present and discuss results[5] obtained for the free energy profiles, mechanisms and rates involved in the nucleation processes of these systems, aiming at understanding the nucleation and growth of crystals of organic salts in the bulk, near surfaces and inside nanopores.

[1] L. Maragliano, A. Fischer, E. Vanden-Eijnden and G. Ciccotti, "String Method in Collective Variables: Minimum Free Energy Paths and Isocommittor Surfaces", *J. Chem. Phys.* **2006**, *125*, 024106

[2] V. Ovchinnikov, M. Karplus and E. Vanden-Eijnden, "Free Energy of Conformational Transition Paths in Biomolecules: The String Method and Its Application to Myosin VI", *J. Chem. Phys.* **2011**, *134*, 085103

[3] L. Maragliano, E. Vanden-Eijnden and B. Roux, "Free Energy and Kinetics of Conformational Transitions

from Voronoi Tesellated Milestoning with Restraining Potentials", *J. Chem. Theory Comput.*, **2009**, *5*, 2589

[4] E. E. Santiso and B. L. Trout, "A General Set of Order Parameters for Molecular Crystals", *J. Chem. Phys.* **2011**, *134*, 064109

[5] X. He, Y. Shen, F. R. Hung and E. E. Santiso, "Molecular Simulation of Homogeneous Nucleation of Crystals of an Ionic Liquid from the Melt", *J. Chem. Phys.* **2015**, *143*, 124506

462117 Improving the Efficiency of Water Adsorption Simulations in Metal-Organic Frameworks

Abstract:

Water is one of the most common components in industrial gas or liquid systems. It often has a non-negligible effect on chemical and physical processes such as gas or liquid adsorption in porous materials, like zeolites and metal-organic frameworks (MOFs). Molecular modeling techniques, especially grand canonical Monte Carlo (GCMC) simulations, have been used by many researchers to study the adsorption of water in porous media and to obtain useful information about the molecular-level interactions. However, due to the high density and complex hydrogen bonding in water systems, these GCMC simulations usually suffer from very low probability of successful insertions, leading to extremely long simulation times especially in the high loading regime. To improve the simulation efficiency for these systems, we selected the hydrophobic MOF ZIF-8 as a representative adsorbent and tested the performance of multiple advanced simulation algorithms, including pre-tabulated energies (â€œgridâ€ method), energy-biased insertions, and continuous fractional component Monte Carlo (CFC MC) for water adsorption. We showed that these methods can improve the insertion and deletion probabilities and the simulation speed compared with conventional GCMC simulation. Furthermore, an alternative computational approach for calculating adsorption isotherms, using molecular dynamics simulations and computation of the chemical potential, was studied and applied to different adsorption systems including water. Some promising results and insights have been discovered for this new approach.

467860 Heterogeneous Ice Nucleation Using Forward Flux Sampling

Abstract:

Rare events are difficult to study in simulations. Rare events are called such because the length and time scales needed to observe the event are significantly longer than the event itself. This means that the expectation time for many rare events exceeds current computational capabilities by many orders of magnitude. One such rare event is ice nucleation. Understanding the role played by solid surfaces in ice nucleation is a significant step toward designing surfaces that can control ice nucleation. Our goal is to explain the mechanisms through which surfaces affect ice nucleation and growth. The challenge in this is to generate statistically significant number of liquid-to-ice transition paths. In our group, we use Forward Flux Sampling (FFS) to address this challenge. However, the adoption of FFS to complex, more realistic, systems requires significant job and data management that can be difficult to do using standard computational scripting tools. We have developed a unique computational framework that combines state-of-the-art Big Data technology with FFS to enable large scale calculations. Using this platform we have studied ice nucleation on silver iodide (AgI) and a modified kaolinite surface. Both these surfaces are a good atmospheric ice nucleating agents and have the same lattice match to ice. Traditionally, lattice match has been used a governing parameter to determine the ice nucleating efficiency of surface. Our extensive calculations show that the rate of ice nucleation on AgI and the modified kaolinite surface is different. In addition, the mechanism of nucleation is also different. This clearly demonstrates that lattice spacing is not a sufficient measure of ice nucleating efficiency. We have also performed committor analysis to identify the reaction coordinates relevant to this transition. To the best of our knowledge, these studies are the first large scale FFS calculations of heterogeneous ice nucleation using all-atom water models.

468923 Theoretical Prediction of Interpenetrating Metal-Organic Frameworks

Abstract:

Theoretical Prediction of Interpenetrating Metal-Organic Frameworks

Kutay B. Sezginel, Tianyi Feng, Christopher E. Wilmer

Chemical and Petroleum Engineering, University of Pittsburgh, Pittsburgh, PA

Metal-organic frameworks (MOFs) are crystalline materials comprising building blocks of inorganic nodes and organic linkers. MOFs offer high permanent porosities, very high surface areas and versatile framework interactions which makes them promising candidates towards various applications such as gas storage [1-3] and separation [4, 5], electrochemistry [6], sensing [7], catalysis [8], and drug delivery [9]. The structural properties of MOFs such as pore size and shape and the chemical environment of the framework can be precisely designed by finding the right combination of building blocks and suitable reaction conditions. An important consideration for the atomically precise design of MOFs is the interpenetration of the framework. Interpenetration is especially observed in MOFs with large void fraction where multiple frameworks interpenetrate each other to form a more thermodynamically favorable and stable product. [10]. In common cases, two, three, or four identical frameworks interpenetrate each other, however, as many as 54 frameworks have been demonstrated to interpenetrate each other [11].

Interpenetration in MOFs if often considered a negative phenomenon since it decreases the available space within the structure. [12] However, interpenetration also provides a degree of freedom in the design of MOFs which results in many beneficial properties such as enhanced framework stability [13], selective gas adsorption [14], increased heat of adsorption [15], adsorption hysteresis [16], photoluminescence control [17] and guest-responsive porosity [18]. Majority of the interpenetrating structures have been discovered as byproduct crystalline phases resulting from MOF synthesis. Therefore, interpenetration is generally observed within the same framework which is also named as *homo-interpenetration.* [19] A much rare case is observed when two distinct frameworks interpenetrate with each other also termed as *hetero-interpenetration.* [20] An exciting prospect in the near future, once control over catenation in MOFs has sufficiently matured, is the selective interpenetration of *different* frameworks that each confer different functionality to the overall material [21] This however, requires a thorough understanding of experimental methods that facilitate control over interpenetration as well as tools that enable rational design of interpenetrated frameworks. [22] Many hetero-interpenetrated MOFs have been reported, with different dimensionalities, [23, 24] topologies, [25, 26] and chemical compositions. [27] Therefore controlling catenation is an important design criterion which has no established control mechanism and remains quite challenging.

Motivated by the promising properties of interpenetrated MOFs and lack of a design tool that enables systematic engineering of these materials, we developed a computational tool to discover interpenetrating MOFs. As a proof of concept, we initially started investigating interpenetration in MOF-5, which is a widely studied homo-interpenetrating cubic MOF. [12] After accurately identifying interpenetration in this MOF we generalized our algorithm to search for interpenetrating structures for any given MOFs. We then investigated a set of experimentally synthesized MOFs from the database constructed by Chung et al. [28] and discovered several hypothetically hetero-interpenetrating MOFs. Our tool is able to discover interpenetration in any given set of periodic network in a matter of minutes (avg. 3.5 min), therefore this method can be easily extended to be used for any periodic crystalline materials.

The interpenetration algorithm starts with creating a potential grid for one of the MOFs. The potential grid is created using Lennard-Jones potential with UFF parameters and a cutoff radius selected according to the size of the MOF unit cell. The potential grid is used to rapidly check for collisions between periodic frameworks and determine the overall potential energy of the structure. After potential grid is constructed for one of the structures, interpenetration of the other structure is checked by screening all the grid points in the potential grid with random orientations by rotating the structure in 3D space. The framework atoms of the interpenetrating network are placed one by one and energy penalty for inserting each atom is calculated. For cases with energy penalty higher than a previously defined energy limit, the interpenetration trial is interrupted and next trial is performed, otherwise insertion of atoms is continued. If all the framework atoms are added without going over the energy penalty, the orientation of the frameworks is recorded. The recorded orientations are checked for collisions in the periodic unit cells and the atomic positions for cases without collisions are exported and sorted according to their overall potential energy. Our initial trials with a set of 50 MOFs (approx. 1250 combinations) resulted in the discovery of tens of interpenetrating frameworks. We are currently screening hundreds of structures to discover novel interpenetrating MOFs. The discovered structures are being investigated in detail and geometrically optimized to identify thermodynamically achievable structures. These interpenetrating MOFs can then be studied in detail to reveal promising properties of these complex structures.

References

1. Sumida, K., et al., *Carbon dioxide capture in metal–organic frameworks.* Chemical reviews, 2011. **112**(2): p. 724-781.

2. Rosi, N.L., et al., *Hydrogen storage in microporous metal-organic frameworks.* Science, 2003. **300**(5622): p. 1127-1129.

3. Eddaoudi, M., et al., *Systematic design of pore size and functionality in isoreticular MOFs and their application in methane storage.* Science, 2002. **295**(5554): p. 469-472.

4. Li, J.-R., R.J. Kuppler, and H.-C. Zhou, *Selective gas adsorption and separation in metal–organic frameworks.* Chemical Society Reviews, 2009. **38**(5): p. 1477-1504.

5. Song, Q., et al., *Zeolitic imidazolate framework (ZIF-8) based polymer nanocomposite membranes for*

gas separation. Energy & Environmental Science, 2012. **5**(8): p. 8359-8369.

6. Xia, W., et al., *Metal–organic frameworks and their derived nanostructures for electrochemical energy storage and conversion.* Energy & Environmental Science, 2015. **8**(7): p. 1837-1866.

7. Kreno, L.E., et al., *Metal–organic framework materials as chemical sensors.* Chemical Reviews, 2011. **112**(2): p. 1105-1125.

8. Lee, J., et al., *Metal–organic framework materials as catalysts.* Chemical Society Reviews, 2009. **38**(5): p. 1450-1459.

9. Orellana-Tavra, C., et al., *Amorphous metal–organic frameworks for drug delivery.* Chemical Communications, 2015. **51**(73): p. 13878-13881.

10. Stock, N. and S. Biswas, *Synthesis of metal-organic frameworks (MOFs): routes to various MOF topologies, morphologies, and composites.* Chemical reviews, 2011. **112**(2): p. 933-969.

11. Wu, H., et al., *An exceptional 54-fold interpenetrated coordination polymer with 103-srs network topology.* Journal of the American Chemical Society, 2011. **133**(30): p. 11406-11409.

12. Kim, H., et al., *Synthesis of phase-pure interpenetrated MOF-5 and its gas sorption properties.* Inorganic chemistry, 2011. **50**(8): p. 3691-3696.

13. Ma, L. and W. Lin, *Unusual interlocking and interpenetration lead to highly porous and robust metal–organic frameworks.* Angewandte Chemie International Edition, 2009. **48**(20): p. 3637-3640.

14. Ma, S., et al., *A Coordinatively Linked Yb Metal–Organic Framework Demonstrates High Thermal Stability and Uncommon Gas-Adsorption Selectivity.* Angewandte Chemie International Edition, 2008. **47**(22): p. 4130-4133.

15. Kesanli, B., et al., *Highly Interpenetrated Metal–Organic Frameworks for Hydrogen Storage.* Angewandte Chemie International Edition, 2005. **44**(1): p. 72-75.

16. Mulfort, K.L., et al., *An interpenetrated framework material with hysteretic CO2 uptake.* Chemistry-A European Journal, 2010. **16**(1): p. 276-281.

17. Takashima, Y., et al., *Molecular decoding using luminescence from an entangled porous framework.* Nature communications, 2011. **2**: p. 168.

18. Maji, T.K., R. Matsuda, and S. Kitagawa, *A flexible interpenetrating coordination framework with a bimodal porous functionality.* Nature materials, 2007. **6**(2): p. 142-148.

19. Carlucci, L., G. Ciani, and D.M. Proserpio, *Parallel and inclined (1D→ 2D) interlacing modes in new polyrotaxane frameworks [M2 (bix) 3 (SO4) 2][M= Zn (II), Cd (II); bix= 1, 4-bis (imidazol-1-ylmethyl) benzene].* Crystal growth & design, 2005. **5**(1): p. 37-39.

20. Zhang, M., et al., *Rational design of metal–organic frameworks with anticipated porosities and functionalities.* CrystEngComm, 2014. **16**(20): p. 4069-4083.

21. Xu, H., et al., *An unprecedented 3D/3D hetero-interpenetrated MOF built from two different nodes, chemical composition, and topology of networks.* CrystEngComm, 2012. **14**(18): p. 5720-5722.

22. Jiang, H.-L., T.A. Makal, and H.-C. Zhou, *Interpenetration control in metal–organic frameworks for functional applications.* Coordination Chemistry Reviews, 2013. **257**(15): p. 2232-2249.

23. Carlucci, L., et al., *Polymeric layers catenated by ribbons of rings in a three-dimensional self-assembled architecture: a nanoporous network with spongelike behavior.* Angewandte Chemie International Edition, 2000. **39**(8): p. 1506-1510.

24. Shin, D.M., et al., *Coordination polymers based on square planar Co (II) node and linear spacer: solvent-dependent pseudo-polymorphism and an unprecedented interpenetrating structure containing both 2D and 3D topological isomers.* Chemical Communications, 2003(9): p. 1036-1037.

25. Qi, Y., Y.-X. Che, and J.-M. Zheng, *Self-penetrating and interpenetrating 3D metal– organic frameworks constructed from the rigid 1, 4-Bis (1-imidazolyl)-benzene ligand and aromatic carboxylate.* Crystal Growth and Design, 2008. **8**(10): p. 3602-3608.

26. Shi, Z.-Q., et al., *Six New 2D or 3D Metal–Organic Frameworks Based on Bithiophene-Containing Ligand and Dicarboxylates: Syntheses, Structures, and Properties.* Crystal Growth & Design, 2013. **13**(7): p. 3078-3086.

27. Lu, Z.-Z., et al., *[WS 4 Cu 3 I 2]– and [WS 4 Cu 4] 2+ secondary building units formed a metal–organic framework: Large tubes in a highly interpenetrated system.* Chemical Communications, 2011. **47**(10): p. 2919-2921.

28. Chung, Y.G., et al., *Computation-Ready, Experimental Metal–Organic Frameworks: A Tool to Enable High-Throughput Screening of Nanoporous Crystals.* Chemistry of Materials, 2014. **26**(21): p. 6185-6192.

470975 First Principles Screening of Transparent Conducting Oxides Using Amobt

Abstract:

Accurate *ab initio* electronic transport models, which bridge electronic and atomistic scales, can facilitate high throughput calculations and screening of semiconductor materials for energy applications. Previous attempts at screening new transparent conducting oxide (TCO) materials have focused on their average effective mass, due to the simplicity and speed of such calculations. This calculation is often simplified by being based solely on the shape of the band extrema. Although the approximate effective mass certainly gives a valuable prediction of material performance, it lacks sufficient complexity for accurate calculation of electronic properties, especially in degenerate semiconductors. In particular, electron-phonon interactions, which limit the mobility, particularly at room temperature, are ignored. Here we employ an *ab initio* transport Model in the Boltzmann Transport (aMoBT) framework, which accurately predicts the electrical mobility and conductivity of both n-type and p-type semiconductors. We calculate electronic structures within the Density Functional Theory framework, and then treat the non-equilibrium behavior of electrons and holes using the atomistic Boltzmann transport equation. We analyze more than 90 promising TCOs that were pre-screened using the average effective mass, and rank them based on their electron and/or hole conductivity, as limited by ionized impurity scattering and electron-phonon scattering mechanisms. We now report the most promising candidates from each of the n- and p-type semiconductors, and assert the utility of aMoBT, as incorporated into an automated atomistic calculation framework, in designing new materials for solar cell applications.

471740 Polymorphic Transformations at Ambient Conditions Elucidated with Molecular Dynamics

Abstract:

A long sought-after goal in the field of computational chemistry is the ability to predict all observable crystal structures, or polymorphs, of small-molecule organic solids. Computer-based models can, in principle, rapidly input a particular molecule and search through possible crystal structures, identifying the set of stable or metastable polymorphs which are likely to be observed experimentally. Current approaches for crystal structure prediction typically rank candidate structures using lattice energies of individual configurations. Such approaches assume that entropy and lattice expansion effects caused by thermal motions at ambient conditions contribute negligibly to the relative stability of different crystal structures. However, many temperature and pressure driven solid-solid transformations are known to exist experimentally, and these cannot be predicted without including thermal motions.

In this work, we present the prediction of finite temperature solid-solid transformations using temperature and pressure dependent free energies derived from a novel molecular dynamics approach. Molecular dynamics simulations implicitly include the thermal motions present at finite temperatures and can therefore predict the emergence of a more stable crystal structure at high temperatures and pressures. We examine the five organic molecules benzene, formamide, glycine, acetic acid, and imidazole which are known experimentally to have different globally stable structures at low and high temperatures. The relative free energies for these solids in the point-charge OPLS potential correctly identify the high temperature / high pressure polymorphs as being more stable than the low temperature form.

We also compute the temperature and pressure dependent free energies in the more expensive polarizable AMOEBA potential to determine the added benefit of using a more accurate energy function. We use a novel Boltzmann reweighting approach to circumvent the additional expense of the polarizable potential which has kept the more accurate model from being used in previous MD studies. The polarizable AMOEBA potential is shown to have a

higher accuracy in predicting the low temperature crystal structure in these systems. However, both OPLS and AMOEBA have equal fidelity at predicting the high temperature and high pressure structure. These preliminary results suggest that high level energy functions are more accurate at predicting low temperature structures, but have a less significant advantage at high temperatures relative to cheaper levels of theory. Finally, we show that these short ambient temperature MD simulations can allow metastable lattice minima to restructure into more stable minima.

473678 NaCl Nucleation from Aqueous Solution By a Seeded Simulation Approach

Abstract:

Nucleation is scientifically and technologically exceptionally important. However, ascertaining mechanisms and kinetics remains extremely difficult for both simulations and experiments even when supposedly simple systems are considered. Conventional rare-event methods typically face a problem related to the nucleus size coordinate n (the measure of nucleation progress). The dynamics of n are so slow that sufficient sampling of the free-energy landscape is not achievable [1]. To overcome this limitation we highlight a seeding approach for molecular dynamics simulations to predict nucleation rates [1] and for ascertaining attachment kinetics [2]. We apply the approach to sodium chloride nucleation from aqueous solution because there are rate measurements for this system in the literature [3–5]. Furthermore, force fields exist that reliably reproduce experimental data on important input data to the approach: the solution chemical potential and the solubility limit [6].

The key idea of the approach is to carefully merge compact clusters cut from a well-equilibrated crystal lattice with a quasi-equilibrated supersaturated solution box. The transient evolution of the nucleus size is then recorded for various seed sizes. The resulting nucleus drift data are then used in the framework of classical nucleation theory (CNT) together with the assumption of over-damped Langevin dynamics for $n(t)$ to estimate the nucleation rate, J. The procedure to calculate J comprises three conceptual steps: (1) estimation of the critical nucleus size from nucleus drift data, (2) determination of the attachment frequency to the critical nucleus from short swarms of trajectories, and (3) fitting of the interfacial free-energy, γ, to reproduce the drift data from step (1).

In step (2), we have tested two models of the attachment kinetics, which reflect diffusion limitation and ion-desolvation limitation, respectively. The diffusion analysis required the calculation of mutual electrolyte diffusivities at varying concentration, which agree well with experimental results from interferometry [7,8].

The diffusion model grossly overestimates the attachment frequency compared with a standard analysis of the mean square change in nucleus size [9]. Therefore, ion-desolvation is identified as the limiting resistance to attachment. A second independent analysis, based on approach-to-coexistence data [10], confirms the negligible influence of diffusion.

Accurate data on the model-inherent solution chemical potential and solubility limit are critical for rate predictions because they define the (thermodynamic) driving force for nucleation in the simulations. The solubility limit of the chosen NaCl–water model has been intensely debated until recently [6,10], for which reason we systematically investigate its influence on the nucleation rate [11]. We find that an error of 30% in the solubility, which reflects the ranges from recent literature [6,10], can cause the rate to differ by ten orders of magnitudes. Our seeded simulations are performed at experimentally relevant concentrations [3]. However, electrolyte chemical potentials, μ, are typically not available at these conditions for the force field chosen. Therefore, we also address the issue of reliably extrapolating μ [11]. Finally, we assess the uncertainty in the rate [11] due to the remaining uncertainty in the solubility [6].

[1] Knott *et al.*, *J. Am. Chem. Soc.* 134, 19544–1954, 2012

[2] Zimmermann *et al.*, *J. Am. Chem. Soc.* 137, 13352–13361, 2015

[3] Na *et al.*, *J. Cryst. Growth* 139, 104–112, 1994

[4] Gao *et al.*, *J. Phys. Chem. B* 111, 10660–10666, 2007

[5] Desarnaud *et al.*, *J. Phys. Chem. Lett.* 5, 890–895, 2014

[6] Nezbeda *et al.*, *Mol. Phys.*, 10.1080/00268976.2016.1165296, 2016

[7] Rard *et al.*, *J. Solution Chem.* 8, 701–716, 1979

[8] Chang *et al.*, *AIChE J.* 31, 890–894, 1985

[9] Auer and Frenkel, *J. Chem. Phys.* 120, 3015–3029, 2004

[10] Aragones *et al.*, *J. Chem. Phys.* 136, 244508, 2012

[11] Zimmermann *et al.*, in preparation

457724 Bottom-up Approach to Constructing a Polymerization Process Model: Leveraging New Insights for Improved Manufacturing Performance

Abstract:

Synthetic polymer materials are commonly used in textiles, automotive applications, carpets, and sportswear due to their high durability and strength. Scale-up and commercial production of these polymer materials is an active research field that has been around for decades and thus spans many types of chemistries. As computational resources are becoming more readily available, new methods are being applied to this broad field to improve manufacturing performance. This presentation summarizes key learnings of a joint effort between process technology and computational modeling groups to provide new insights to aid process improvements. Detailed discussion will be focused on how computational modeling efforts supported by experimental kinetics studies have been leveraged to contribute to process development. At the atomistic level, Density Functional Theory (DFT) was employed to deliver insights into the underlying reaction kinetics and thermochemical properties. Also, kinetic model building and validation schemes will be explained to demonstrate how first-principles-based quantum chemical calculations have been incorporated to construct a predictive process model for polymer chemistry.

451217 A Computational Tool-Box for Pharmaceutical Chemical Development

Abstract:

Pharmaceutical Drug Development branch faces challenges of accelerated development under a high regulatory pressure. An ability to rationalize and guide drug development process has become of crucial importance [1,2]. A deep integration of computational modeling into Drug Development is required to achieve this goal. State-of-the art computational approaches were created based on Pipeline Pilot Web Port architecture, to support process and analytical chemistry development in the pharmaceutical industry. These computational chemistry tools include multiple approaches related to reactivity and free energy of reaction prediction to guide chemistry route analysis and selection problems. Selected applications of these computational tools will be presented.

1.Ouyang, D.; Smith, S. C., Eds. *Computational Pharmaceutics: Application of Molecular Modeling in Drug Delivery*, John Wiley and Sons, 2015.

2. Abramov, Y.A. , Ed. *Computational Pharmaceutical Solid State Chemistry*, John Wiley & Sons, 2016.

465105 A Critical Evaluation of the Accuracy of Thermoset Models Created Using Atomistic Simulation Methods, Based on Gelation Studies

Abstract:

In recent years, a steadily increasing number of molecular modeling studies have focused on prediction of mechanical and thermophysical property prediction for thermosetting polymeric materials, driven in large part by incorporation of such materials in advanced composites used in aerospace, automotive and electronics applications. Although comparison of property predictions has shown encouraging results (e.g. the successful and quantitative prediction of the effect of epoxy resin precursor molecular architecture on small strain mechanical behavior [1]), detailed examinations of the topological and chemical accuracy of the molecular models used have thus far been limited.

Since model accuracy must ultimately determine the ability of molecular simulations to predict a wide range of properties relevant to industrial applications, we have recently concentrated on extending earlier studies to include more detailed evaluation of model quality for common thermosetting materials. Rather than relying on comparisons on calculated and measured values of properties related *indirectly*to network structure (e.g. as is the case with elastic constants), we instead choose to focus on the experimentally observable gel point, which has long been recognized both theoretically and experimentally to depend on chemical functionality of reactive groups and on aspects of the crosslinking process, such as cyclization, which determine when incipient infinite network formation occurs [2].

In this presentation we begin by summarizing both the experimental methods used to measure the gel point together with published experimental data for a variety of network-forming reactions, before proceeding with details of the simulations and gel point predictions obtained therefrom. The types of systems studied in the simulations include the popular amine-cured epoxy systems in addition to polyester containing materials. For selected systems, we consider possible artifacts associated with the relatively small size of the models, and other factors such as the customary neglect of less frequent secondary reactions or unequal reactivities of the various reacting groups. Finally, while for the most part the studies have focused on so-called RA_{fa}+$R'B_{fb}$ systems formed by a step growth polymerization mechanism (with fa and fb referring to chemical functionalities of A and B groups), we will briefly consider networks formed by chain growth polymerization.

References

1. Rigby, D, Saxe, P.W., Freeman, C.M. and Leblanc, B., "Computational Prediction of Mechanical Properties of Glassy Polymer Blends and Thermosets", *in*"Advanced Composites for Aerospace, Marine, and Land Applications", T. Sano, T.S. Srivatsan and M.W. Paretti, eds., Ch. 14, John Wiley & Sons (2014).

2. *see, for example*, Flory, P.J. "Principles of Polymer Chemistry", Ch. 9, Cornell University Press, Ithaca, New York (1953).

466486 Molecular Simulation of *N*-Octacosane – Water Gtl Mixtures in Nanopores at Elevated Temperature and Pressure

Abstract:

Water - wax mixtures are products of the Gas-To-Liquids (GTL) process through the Fischer-Tropsch (FT) route, a polymerization reaction converting syngas (H_2 and CO)[1] into liquid hydrocarbons and, to a smaller extent, oxygenates (e.g. alcohols). Transport properties of these mixtures, particularly inside catalyst nanopores, is a topic of significant interest for the petrochemical industry. The choice of material for catalyst support is of paramount importance as its surface structure and pore size,[2] is essential in ensuring FT reactor's activity and stability, given that these features control metal catalyst dispersion, hydrocarbon selectivity reactant and product diffusion.[3] Typical catalyst carriers are alumina, silica, titania, carbon, etc.[4] In this study, we chose titanium dioxide as it is commonly used in the literature and its structure is well defined.[5] It is generally assumed that inside the catalyst nanopores and under reaction conditions, the overall mole fraction of water inside the nanopores varies from less than 0.01 up to as high as 0.20, corresponding to the top and the bottom, respectively, of the FT fixed-bed reactor. This assumption may not hold especially in the outlet of the FT reactor, where water concentration is significantly higher.[6] Furthermore, excess water inside the nanopore could lead to sintering of catalytic nanoparticles,[7] shortening the catalyst's lifetime and increasing GTL plant's operational costs. It is thus important to understand: a) the water–wax vapor–liquid equilibrium (VLE) at FTS reactor operation conditions, as it determines the maximum allowable amount of water in the FT wax, b) the phase behavior of water–wax mixture inside the nanopores, by incorporating wall effects and c) the role of oxygenates (e.g., alcohols and acids < 10 wt % in wax)[1] on the phase behavior of water–wax inside nanopores and their interactions with the catalyst wall.

The present study focuses on simulating the phase behavior of the *n*-octacosane (n-C_{28}) – water mixture

inside TiO_2 nanopores. Molecular Dynamics (MD) simulations with realistic molecular models[8-11] were employed, in order to account for the fluid–fluid and fluid–pore interactions, pore shape and size. Our simulations reveal the importance of confinement especially on the excess water–wax mixture transport properties, as water molecules tend to organize into two discrete layers on the TiO_2 surface.

Keywords: Gas-to-Liquids, water/*n*-octacosane mixture, Molecular dynamics

References

1. Wood, D. A.; Nwaoha, C.; Towler, B. F., *J. Nat. Gas Sci. Eng.* **2012,** *9*, 196-208.

2. Khodakov, A. Y.; Chu, W.; Fongarland, P., *Chem. Rev.* **2007,** *107*, 1692-1744.

3. Jahangiri, H.; Bennett, J.; Mahjoubi, P.; Wilson, K.; Gu, S., *Catal. Sci. Technol.* **2014,** *4*, 2210-2229.

4. Oh, J.-H.; Bae, J. W.; Park, S.-J.; Khanna, P. K.; Jun, K.-W., *Catal. Lett.* **2009,** *130*, 403-409.

5. Jongsomjit, B.; Sakdamnuson, C.; Praserthdam, P., *Mater. Chem. Phys.* **2005,** *89*, 395-401.

6. Panahi, M.; Rafiee, A.; Skogestad, S.; Hillestad, M., *Ind. Eng. Chem. Res.* **2012,** *51*, 425-433.

7. Qin, Q.; Ramkrishna, D., *Ind. Eng. Chem. Res.* **2004,** *43*, 2912-2921.

8. Martin, M. G.; Siepmann, J. I., *J. Phys. Chem. B* **1998,** *102*, 2569-2577.

9. Utesch, T.; Daminelli, G.; Mroginski, M. A., *Langmuir* **2011,** *27*, 13144-53.

10. Bandura, A. V.; Sykes, D. G.; Kubicki, J. D., *Abstr. Pap. Am. Chem. S.* **2002,** *223*, U603-U603.

11. Berendsen, H. J. C.; Grigera, J. R.; Straatsma, T. P., *J. Phys. Chem.* **1987,** *91*, 6269-6271.

451176 Mind the Gap: Bridging Molecular and Fluid Dynamics Simulations in the Polyurethane Foam Industry

Abstract:

In a previous work, a novel approach, based on the incorporation of Computational Fluid Dynamics (CFD) with the Population Balance Equation (PBE) was developed and applied for the first time to the simulation of expanding polyurethane foams. The model included a simple but realistic representation of the polymerization kinetics. The model also accounted for the presence of a physical blowing agent and of a chemical blowing agent. Notwithstanding its simplicity, a good agreement with experimental data was obtained. Yet, some limiting assumptions were adopted in the development of this approach, as follows:

1) the density of the polymerizing liquid mixture, assumed fixed to a constant value and, as such, independent of temperature and degree of cross-linking;

2) the temperature dependence of the solubility of the physical blowing agents (e.g., R-11, in the case of physically blown foams) and of the produced carbon dioxide (in the case of chemical blown foams).

To fill this gap in the present work we present an implemented version of this approach according to which molecular simulations are employed to calculate the missing quantities at points 1) and 2), and their link to the CFD code for the more realistic simulation of expanding physically and chemically blown foams.

This work was funded by the European Commission under the grant agreement number 604271 (Project acronym: MoDeNa; project identifier: FP7-NMP-2013-SMALL.

455569 How Uncertainty Quantification Renders Molecular Simulation a Quantitative Tool for Thermophysical Property Evaluation

Abstract:

The critical temperature (T_c), critical pressure (P_c), and critical density (ρ_c) are important thermophysical properties in science and engineering. The critical constants are required parameters in many prediction models for other thermophysical properties, such as saturated liquid density, liquid heat capacity, second virial coefficient, liquid thermal conductivity, liquid vapor pressure, and liquid viscosity. As reliable estimates of thermophysical properties are essential for engineering design purposes, the Design Institute for Physical Properties (DIPPR 801) desires accurate predictions of T_c, P_c, and ρ_c. The n-alkane family is of particular interest because it is traditionally assumed that the critical constants for a homologous series converge to the n-alkane family with increasing chain-length. Unfortunately, obtaining reliable critical constant values becomes difficult for larger compounds that thermally decompose at temperatures below T_c.

For these reasons, many researchers have attempted to develop prediction models and sophisticated experimental approaches to estimate the critical constants for large n-alkanes. However, the different prediction models and experimental data sets differ considerably for larger compounds. Furthermore, prediction models can lead to erroneous results when extrapolating to chain-lengths that are longer than the data used in model development. In addition, experimental T_c, P_c, and ρ_c values for large n-alkanes are subject to a great degree of uncertainty. Molecular simulation provides a viable approach to discern between these conflicting and dubious results and, thereby, elucidate the correct trends for T_c, P_c, and ρ_c with respect to carbon number for large n-alkanes.

A significant factor when assessing the reliability of any physical property value, obtained either from experiment, model prediction, or molecular simulation is assigning the uncertainty associated with that value. Uncertainty quantification and propagation of error is, therefore, an essential tool for industrial process design. Unfortunately, determining an appropriate uncertainty is normally more straightforward for experimental measurements than for molecular simulations. This is because molecular simulations have many layers of uncertainty that propagate and are difficult to quantify. These layers of uncertainty can be listed in a bottom-up order as: the quality and type of data used for developing the model (liquid density, vapor pressure, ab initio energy calculations, spectroscopic data, etc.), the intramolecular model used (rigid bond lengths, flexible bonds, angles, etc.), the type of site model chosen (all-atom, united-atom, anisotropic-united-atom, etc.), the mathematical form of the model chosen (Lennard-Jones, Exponential-6, etc.), the optimization of the model parameters (ϵ, σ, q, etc.), the simulation limitations (finite-size effects, cutoff errors, pairwise additivity, etc.), the type of simulation (GEMC, GCMC, MD, etc.), the analysis of simulation data to yield the desired property (rectilinear diameters and density scaling for T_c and ρ_c, Clapeyron equation for P_c, etc.), and the experimental design conditions (temperature, composition, etc.). These layers are typically categorized into three different types of uncertainty, namely, model form, model parameterization, and numerical uncertainties.

Although most molecular simulation studies of large compounds report small uncertainties compared to the experimental uncertainties, these typically represent only the numerical uncertainty. In this study we have implemented a systematic investigation of several layers of uncertainty when predicting the critical constants of large n-alkanes. By so doing, we have enhanced the ability of molecular simulation to be a quantitative tool in physical property assessment and process design. Specifically, we obtained the first quantitatively reliable estimates of P_c for large n-alkanes. This allowed us to determine which set of experimental data are most reliable for large compounds. After performing this analysis we optimized a new prediction model for P_c. Furthermore, by obtaining reliable estimates for T_c and ρ_c from molecular simulation we developed new prediction models for T_c and ρ_c. The T_c and P_c models enabled the development of a normal boiling point (T_b) model by ensuring that $T_c = T_b$ at the same carbon number that $P_c = 1$ atmosphere. Therefore, the quantification and reduction of molecular simulation uncertainties were influential in developing internally consistent prediction models for T_c, P_c, ρ_c, and T_b for n-alkanes. These models enable the prediction of many other important thermophysical properties for process design.

- Abstract Text

468955 Alpha and Beta Glycine Nanocrystal Growth and Dissolution Kinetics in the Presence of an Electric Field through Molecular Dynamics - Towards Electric Field Controlled Crystallization

Abstract:

Control of crystal polymorphs is of significant interest in the field of drug discovery and development. An active pharmaceutical ingredient (API) can exhibit different crystal forms, which in turn determine its solid-state stability, bioavailability, dissolution rate, as well as other key solid-state characteristics. It is well known that while a particular polymorph may be more thermodynamically stable, its rate of formation (kinetics) may be slower relative to its metastable counterpart. Consequently, a pharmaceutical crystallization process that has been designed to produce only the desired stable polymorph may result in different, or worse yet, undiscovered, polymorphic forms, due to unforeseen variations in the manufacturing process conditions. It is therefore essential to not only understand, but also control, the mechanisms of polymorph formation. This includes understanding of not only the competitive nucleation and growth process, but also dissolution as one attempts to selectively dissolve potentially unwanted polymorphs in favor of the desired forms. Ultimately, this mechanistic understanding must translate to quantitative estimates to be used in design and control models. Recent advancements of ultrafast laser physics have shown that intense, pulsed laser is highly effective in manipulating various molecular states in the non-resonant regime. Specifically, several experiments have illustrated the effect of continuous wave (CW) as well as pulsed laser in a crystallization system. These include accelerated crystallization of urea, crystallization of glycine in undersaturated solution, and glycine polymorph selection. As the atomic forces due to the application of an external field can be represented classically via Lorentz Law in the molecular Hamiltonian, electric field controlled crystallization presents the tantalizing possibility of placing the experimentalist in direct control of the underlying free energy surface upon which the crystallization process is taking place. This manipulation of free energy barriers will manifest itself in nucleation, growth, and dissolution electric field dependent rates. As a step further towards electric field controlled crystallization, we seek to determine, using molecular dynamics (MD), first principle, and polymorph specific, dissolution and growth rates for various electric field intensities. We believe this to be an important step towards the ultimate goal, which is expressing the growth, and dissolution, rates as functions of the applied electric field.

470886 Cheminformatic Selection of a Fluid for Use in a Single-Loop Thermal Control System

Abstract:

The heat transfer fluid (HTF) for a single-loop thermal control system is selected according to the temperature of heat rejection. Additionally, the application can impose many different restrictions on the HTF such as toxicity, materials compatibility, pour point, boiling point, or flammability. Mainstream Engineering used cheminformatic techniques to identify promising new heat transfer fluids with good materials compatibility, low toxicity, and optimized thermal performance. Data from chemicals with known properties were used to generate predictive models of fluid properties. The properties of interest to this study were thermal conductivity, viscosity, heat capacity, density, boiling point, flash point, and melting point. The models were optimized to best predict larger and more complex molecules which had a wide operating temperature range. The chemicals predicted to be the best performers were then further evaluated on toxicity and flammability bases. The potential HTFs were experimentally evaluated to validate the model predictions. Model development and property testing as part of a single-loop thermal control system design strategy will be presented.

453470 Comparative Analysis of Biomimetic CO_2 Hydration over Computationally Designed Ruthenium Cluster Decorated Graphene/Carbon Nanotube

Abstract:

Ruthenium cluster decorated graphene (Ru/Gr) and carbon nanotube (Ru/CNT) were optimized to observe the catalytic activity of CO_2 hydration reaction over the proposed catalysts using density functional theory calculations. Ru/Gr and Ru/CNT were found to be active for biomimetic CO_2 hydration reaction. CO_2 hydration mechanism over Ru/Gr and Ru/CNT followed the CO_2 hydration mechanism catalyzed by α-carbonic anhydrase enzyme. For CO_2 to HCO_3^- ion conversion, four elementary steps were observed as OH^- ion formation with H_2O dissociation, CO_2 bending in presence of OH^- ion, HCO_3^- ion formation by proton migration and HCO_3^- ion desorption with H_2O addition. HCO_3^- ion formation step was found to be the rate limiting step. Potential energy surfaces were developed for the reaction. Activation barrier of H_2O dissociation was observed lesser over Ru/CNT than Ru-Gr. Ru/CNT was found to be more active than Ru/Gr.

478499 Impact Award: Inverse Design of Interactions for Assembly

Abstract:

Nanometer-scale, colloidally-stable particles suspended in a fluid can be driven to assemble into a wide variety of different structures depending on the control parameters of the system and the nature of the effective interparticle interactions. In many cases, the relevant interactions are tunable via external fields, physical or chemical modification of the particle surfaces, or changes in the composition of the suspending solvent. In this talk, we discuss some of the theoretical challenges associated with the inverse design of interactions for assembly into a targeted structure and the opportunities that new

machine learning based simulation approaches provide for addressing them.

478501 Predicting Diffusion and Viscous Dissipation in Therapeutic Antibody Solutions

Abstract:

Therapeutic antibodies represent the fastest growing segment of the biopharmaceutical industry. Their application to treatment of cancers and autoimmune diseases is prolific. The engineering of antibodies, processes for their production, and mechanisms for administration is of intense scientific interest. Because of their sequence specific, heterogeneous charge distribution, antibodies can interact strongly, which has a significant impact on solution rheology. The viscosity of antibody solutions is a key parameter and constraint on processes such as ultra-filtration and administration routes such as subcutaneous or intravitreal injection. No methods exist currently to predict the rheology and dynamics of concentrated antibody solutions. I will present the results of a study of two antibodies via coarse-grained simulations. The simulations incorporate hydrodynamic interactions between the suspended proteins as well as appropriate colloidal interactions among the coarse-grained domains. The simulations reproduce the static structure factor of each antibody measured via neutron scattering, and different estimates of the intermolecular friction are used bound the short-time self-diffusivity measured via neutron spin echo. The shear stress autocorrelation function and the zero shear viscosity are computed. Good agreement is found with the measured viscosity of one of the antibodies. However, the simulations significantly underestimate the viscosity of the other antibody. This too can be attributed to uncertainty in the intermolecular friction and suggests potential avenues for future investigation.

480439 Characterizing Hydration at Nanostructured Surfaces – Applications to Materials Design and Protein Interaction Prediction

Abstract:

By understanding the molecular underpinnings of solvation (of water and other liquids) adjacent to complex nanostructured surfaces, that is, surfaces with chemical patterns and/or texture at the nanoscale, our group strives to inform the design of the next generation of advanced materials, and to facilitate efficient and accurate prediction of biomolecular interactions and assembly. In this presentation, I will provide examples of our recent work in both of these areas.

I will first highlight our efforts towards the rational design of superhydrophobic surfaces, that is, hydrophobic surfaces, which utilize roughness to impede the wetting of the surface. While such surfaces display numerous desirable properties, including water-repellency, self-cleaning, and interfacial slip, a key bottleneck in the widespread adoption has been the irreversible loss of superhydrophobicity, which accompanies the wetting of the surface texture at elevated pressures. Using molecular dynamics simulations in conjunction with enhanced sampling techniques, we discovered that water density fluctuations play a crucial role in the dewetting of the surface texture, stabilizing a non-classical dewetting pathway, which involves a number of transitions between distinct dewetted morphologies, and lower free energetic barriers to dewetting. Importantly, we were able to use these insights to augment the surface texture, and demonstrate for the first time, that the barriers to dewetting on textured surfaces could be eliminated altogether, thereby allowing the surface to spontaneously recover its superhydrophobicity.

I will then describe our efforts in uncovering the role that water plays in mediating the interactions and self-assembly of complex molecules, including proteins, peptides, and surfactants. The extent to which the

inherent structure of water is perturbed by these complex molecules, determines the thermodynamics and the kinetics of their assembly. However, accurately characterizing perturbation is challenging, because proteins have incredibly complex surfaces that disrupt the inherent structure of water in countless different ways, depending not only on the chemistry of the underlying protein surface, but also on the precise topography and chemical pattern of amino acids. I will discuss our recent efforts on quantitatively characterizing this disruption of water structure, with the goal of using this information to efficiently predict the interfaces through which these complex monomers interact with one another and self-assemble.

478503 Young Investigator Award: Statistical Learning of Viral Fitness Landscapes for *in silico* Vaccine Design

Abstract:

The coupling of “big computing” – petascale systems and the multicore paradigm – with “big biology” – high-throughput sequencing – present new opportunities for statistical, molecular, and data driven modeling of biological systems. We have developed an approach rooted in spin glass models from statistical physics to translate clinical sequence databases of hepatitis C virus into quantitative molecular models of viral fitness. Mapping out the replicative fitness of the virus as a function of its amino acid sequence, these empirical fitness landscapes prescribe the mutational playing field over which the virus is constrained to evolve. Exploiting the isomorphism of these landscapes to free energy surfaces, we have identified particular combinations of mutations that can induce a phase transition and collapse of the viral population. Agent-based predator-prey simulations of the viral dynamics over the fitness landscape under host immune pressure can identify immune responses capable of crippling viral fitness. Together these approaches reveal vulnerable immune targets in the viral proteome for the rational design of hepatitis C vaccines to hit the virus where it hurts.

478504 Computationally Connecting Organic Photovoltaic Performance to Atomistic Arrangements in Bulk Morphologies

Abstract:

Abstract: In order to meet projected global energy demands over the next 25 years, the equivalent of building a 2 GW power plant every few days is needed. Existing clean power generation technologies can meet this demand in principle, but their relatively large short-term costs limit their widespread adoption. In this work we aim to enable manufacturing of organic solar panels that overcome economic barriers to adoption by optimizing the structure of the organic active layer responsible for generating electricity. We perform coarse-grained molecular dynamics simulations accelerated with graphics processing units to determine the thermodynamically stable morphologies for a variety of candidate ingredients. Using these simulations we explain how morphology can be controlled by molecular self-assembly, identify promising candidate ingredients, and perform the first calculations incorporating both large-scale morphologies and first-principles calculations to determine device-scale charge mobilities.

467553 Will Molecular Modeling Ever Become a Mainstream Chemical Engineering Tool?

Abstract:

Most people have heard the quip "Everyone trusts the results of an experiment except the person who did it. Nobody trusts the results of a molecular modeling study except the person who did it."

While funny, this statement conveys several important truths. First of all, performing careful experiments is difficult and those who are the most careful are also the most skeptical of their own measurements. Good experimentalists know how many things can go wrong, and so when it comes to experimental research, we have come to insist on reproducibility and transparency. By and large the problems of reproducibility and transparency have been solved for many "mainstream" chemical engineering experimental tools such that complex techniques like NMR, X-ray diffraction, calorimetry, and chromatography are used routinely. Results from these techniques are believable because the instruments used in such experiments are commercially available and the databases and software analysis tools used to interpret the output from these instruments are well established and validated. Anyone with a similar instrument who follows the description of how the experiment was conducted could replicate the experiment. Such reproducibility is one of the main principles of the scientific method and gives people confidence that the results are "right".

If we are honest, molecular modeling is not yet close to being a trusted, reliable and reproducible tool, which is the second thing the quip points out. There is still widespread distrust of the veracity of molecular modeling work among many experimentalists. Their skepticism is justified, in my view, and usually has to do with concerns about the use of classical potentials or small systems or short time scales. While these are all valid areas of concerns, molecular modeling practitioners know that there are many other things to worry about. Is there an error in the force field parameters? Was the proper cutoff used? Has the system been properly equilibrated? Was a valid analysis carried out in order to arrive at a property? Is there a bug in the code? It is these latter areas of concern that we as a community can and must address if molecular modeling is ever going to be a mainstream tool.

In this talk, I will highlight what I think some of the biggest challenges are when it comes to making molecular modeling more accurate, reliable and reproducible. I will emphasize best practices that we as a community should adopt and point out some examples of initiatives that are going on that are moving us in the right direction.

467478 Expanding Molecular Simulation Use By Data/Code Sharing in Scientific Publishing

Abstract:

Molecular simulations hold a unique property in research methodologies; in the vast majority of cases the results can be exactly reproduced if the input and codes are known. Despite this, it is rare to find fully reproducible reports of molecular simulations in the literature. Especially for new-comers to the field, this makes it difficult for researchers to learn from, reproduce and build on results from the literature. Data sharing and reproducibility of research are increasingly important issues. Funding agencies are mandating data sharing in calls for proposals, and journals are increasingly requiring data sharing as a condition of publication. Scientists are increasingly interested in open access to data. A requirement, or even desire to share is not sufficient, however, if sharing is difficult or tedious. We believe that new authoring tools are needed that will integrate data and analysis into the research and publishing processes. These tools will reduce the difficulty of sharing and reusing data.

We have developed a new approach to writing scientific documents that enables the direct inclusion of human-readable, and machine-addressable data and code in the narrative text. In this talk we will illustrate the approach by example from papers [1,2] we have recently published using the approach. We will show examples of how we use this approach to document molecular simulation results, and how other researchers can see what we have done, and reuse the code/data from the published manuscripts for new purposes. We show that the combination of an extensible editor (Emacs) with a lightweight markup language (org-mode) provides a remarkable solution to data sharing and research reproducibility issues. This combination enables the documentation of experimental setup, data generation, and analysis in a single document, and subsequent export of a scientific manuscript that is suitable for submission to most journals. When coupled with external data repositories, the approach enables sharing of large or complex data sets that cannot easily be captured in a manuscript.

1. Kitchin, John R., Data Sharing in Surface Science, Surface Science, in press (2015). doi:10.1016/j.susc.2015.05.007.
2. Kitchin, John R., Examples of Effective Data Sharing in Scientific Publishing, ACS Catalysis, 5(6), pp. 3894-3899 (2015). doi:10.1021/acscatal.5b00538.

470979 Integrated Upstream and Downstream Data Curation Tools As a Key to Enabling Reproducibility, Usability and Data Sharing

Abstract:

In this presentation, we describe the development of a computational â€œworkbenchâ€ whose goal is to provide an integrated computational and data environment to support multiscale modeling of soft materials for the Materials Genome Initiative (MGI). The design has three essential elements: a modular program structure that supports the addition of new functionality through Python scripting and run-time plugins; a hierarchical data structure which enables unified representation of materials at different levels of granularity; finally, integration of the NIST Materials Data Curation System (MDCS) [1-2] into the environment to support ontology based materials descriptions. A key element of the design which we emphasize in this presentation is the database element. The XML schema based database environment allows us to visualize the inter-relationships between data elements, and enables automated curation of both upstream and downstream data in the workflow.Â We show how controlling the data in this manner is essential for ensuring reproducibility, results in greatly enhanced usability, and allows users to build progressive, materials reference libraries which can be pushed or shared by various means. We will illustrate this using various examples including tools being developed for coarse-grained force-field development and property calculation tools.

References

1. Materials Data and Informatics, http://www.nist.gov/itl/ssd/is/materials-data-and-informatics.cfm
2. Materials Data Curation System, https://github.com/usnistgov/MDCS

467424 High-Throughput Calculations of Molecular Properties in the MedeA ® Environment

Abstract:

The reliable and efficient determination of chemical and physical properties of large sets of organic and inorganic molecules is of high industrial value for chemical engineers, yet it still remains a formidable scientific and methodological challenge [1]. The requirements of the European Union Registration, Evaluation, Authorization, and Restriction of CHemical substances (EU-REACH) and the Toxic Chemicals Safety Act in USA (US-TCSA) stress the importance of this capability. In chemical engineering, the thermodynamic property predictions of pure organic compounds are mostly relying on group contribution methods, corresponding states theory, and equations of state. Beyond this, molecular simulations offer an alternative, but have hitherto been considered as too computationally demanding and not sufficiently flexible to provide timely predictions to engineers. With the progress in compute power and algorithmic developments, however, computer simulations can now play a role in fulfilling the objectives of evaluation of the physical and chemical properties of thousands of chemical compounds. In particular, computer simulations permit the standardization and automation of tasks that can then be spread over large numbers of processors resulting in highly parallelized computations.

The atomistic and molecular simulation environment MedeA® [2] is capable of handling the calculations of molecular and materials properties for batches of thousands of different structures. We illustrate this with the determination of the accuracy of the semiempirical package MOPAC2012 [3] to compute thermodynamic properties for 1,400 organic and inorganic molecules. The accuracy is validated by comparing with experimental data and density functional theory (DFT) values. The determination of other thermochemical and thermophysical properties of fluids and solids obtained using Monte-Carlo simulations with Gibbs [4], molecular dynamics simulations with LAMMPS [5], and periodic DFT simulations with VASP [6] is also discussed.

References

[1] C. Nieto-Draghi, G. Fayet, B. Creton, X. Rozanska, P. Rotureau, J.-C. de Hemptinne,P. Ungerer, B. Rousseau, C. Adamo, Chem. Rev. 2015, 115, 13093-13164.

[2] MedeA®: Materials Exploration and Design Analysis Copyright © 1998-2014 Materials Design, Inc. Version 2.14.6 (http://www.materialsdesign.com).

[3] MOPAC2012, James J. P. Stewart, Stewart Computational Chemistry, Colorado Springs, CO, USA, http://OpenMOPAC.net (2012).

[4] Gibbs 9.3, IFP Energies Nouvelles, Rueil-Malmaison, Laboratoire de Chimie-Physique, Université de Paris Sud, and CNRS.

[5] S. Plimpton, J. Comput. Phys., 1995, 117, 1 (http://lammps.sandia.gov).

[6] G. Kresse, J. Hafner, Phys. Rev. B, 1993, 47, 558.

470371 Application of Concepts from Modeling Integrated Computing for Molecular Simulation for Workflow Encapsulation

Abstract:

Over the last few decades, the molecular simulation community has worked together to develop a host of open-source simulation packages, as well as many transferable force fields, substantially reducing the barrier for performing simulations. However, despite these efforts, much of a simulation workflow (i.e., the steps required to run and analyze simulations) are still performed using ad hoc codes, scripts and procedures, limiting the reproducibility of simulation results by other researchers, and making it challenging to generally perform property screening and optimization. To translate these complex workflows into a set of tasks that can be automated, combined with other tasks (such analysis routines) and embedded within hierarchies of application-hardened engineering methodologies (such as stochastic optimization), we borrow concepts from the computer science field of model integrated computing (MIC). MIC addresses the problems of designing, creating, and evolving information systems by providing rich, domain-specific modeling environments and tools for model analysis and synthesis. Here, we report our efforts to develop tools to facilitate system initialization, atom-typing, and workflow design that integrate with the ever-growing scientific Python stack. These tools and procedures help to capture the details required to retain knowledge of the tasks and processes used to perform a simulation; this augments reproducibility, aids in training new researchers to perform complex simulation operations, and ultimately enables general screening operations to be performed.

468025 Making Molecular Modeling and Simulation a Mainstream Source of Data for Chemical Engineering Applications

Abstract:

The Journal of Chemical Engineering and Data (JCED) is spearheading an initiative aimed at making molecular modeling and simulation (MMS) a mainstream source of data for chemical engineering applications. To increase acceptance of MMS data in the community requires additional efforts in ensuring reproducibility, providing detailed statistical analysis, assessing accuracy, specifying and verifying software and algorithms, and validating molecular models and force fields. This talk will describe how JCED is advancing these goals by developing and enforcing requirements for MMS data to be suitable for publication in the Journal.

468691 The NIST Standard Reference Simulation Website: Reference Calculations and Evaluated Thermodynamics Properties to Aid Molecular Simulation Users

Abstract:

Molecular simulations, in particular Monte Carlo Molecular Simulation and Molecular Dynamics Simulation, are methods for computing the properties of some kind of system, for which the intermolecular interactions are specified, via mathematical relationships derived from statistical mechanics. Since statistical mechanics itself makes no reference to molecular simulation of any kind, the properties that derive from statistical mechanics should not depend on the particular simulation algorithm used to compute properties for a specified model. Consequently, the results obtained from a molecular simulation should be characteristic of the of the model system, subject only to statistical uncertainty, provided that the simulation technique follows necessary rules (e.g., detailed balance, microscopic reversibility, etc.) and is computationally robust. Standardized results for molecular simulations are, then, achievable. Yet, there are few examples of standard molecular simulation results available to users. Furthermore, documentation of molecular simulations in literature articles is often poor or incomplete. In this paper, we discuss efforts at NIST to produce the â €œNIST Standard Reference Simulation Website.â€[1] The goal of this ongoing project is to publish well-documented simulation results for reliable simulation software for a variety of systems, many of which are commonly used in molecular simulations, that can be treated similar to "standard reference data" that has been generated in an laboratory setting and critically evaluated. The data provided by this NIST project may be used to validate and test new or existing molecular simulation software, as reference data, or for other appropriate applications.

[1] Shen, V.K., Siderius, D.W., and Krekelberg, W.P., Eds., NIST Standard Reference Simulation Website, NIST Standard Reference Database Number 173, National Institute of Standards and Technology, Gaithersburg MD, 20899, http://www.nist.gov/mml/csd/informatics_research/srsw.cfm

471715 What's the Molecular Simulation Checklist?Â Automating Both the Simulation and Research Choices to Achieve Better Robustness and Reproducibility

Abstract:

A recent push in many fields requiring expert knowledge of complex procedures has been to develop checklists for these procedures encoding best practices. For example, airplane pilots use checklists for most flight tasks, and checklists for preventing central line infections in hospitals have resulted in reductions in these infections by more than half in many studies.

What might be the molecular simulation best practices checklists? I discuss efforts to develop such checklists. Some such checklists would be used for validating software and simulation procedures. For example, simple automated analysis procedures can allow statistical determination of proper sampling according to a desired thermodynamic ensemble, as well as identifying ranges of input parameters that give results inconsistent with the expected results. Other checklists should be be aimed at practitioners, identifying what choices to make (or not make) to maximize the chance of yielding reproducible results.

457251 Computational Study on the Thermal Degradation Mechanism and Gas Adsorption Properties of Mesoporous Silica MCM-41 after High Temperature Treatment

Abstract:

Computational modeling is a powerful tool to investigate high temperature processes that are not experimentally facile. In particular, MCM-41-type mesoporous silica, which has potential applications in the capture and storage of gases, was studied here with both Molecular Dynamics (MD) simulation and Monte Carlo (MC) simulation in order to investigate the porous structure transformation under high-temperature treatment up to 5000 K, and its resulting gas adsorption/separation properties. It was found that the process starts with the breakage of Si-O bonds in the solid matrix, followed by reformation at the surface, ultimately leading to pore collapse. This degradation process is dependent on several factors including heating rate, such that complete pore closure is postponed at high heating rates as simulated by MD. By applying the Kissinger equation, a strong correlation between simulated pore collapse temperature and experimental values was found and yielded activation energy of 373.8 kJ/mol for the pore closure. Monte Carlo simulation was then applied to better understand how the original MCM-41 and post-thermally treated structure (2650 K) behave when adsorbing/storing different noble gases at 300 K for pressures varying from 1 kPa to 10000 kPa. Our study revealed the rougher surface and smaller pore size of the 2650 K model can increase the adsorbate-adsorbent interaction due to a curvature effect, and the phenomenon is amplified with the increase of gas molecule size. The selectivity between two different noble gases was then found to be improved by the thermally treated model.

453050 Methane-Hydrate Nucleation in Marine Environments: Insights from Molecular-Dynamics Strategies Exploring Free-Energy Landscapes

Abstract:

Molecular-dynamics simulations of methane hydrates in contact with silica surfaces have been performed, mimicking the marine environment. A set of suitable collective variables was employed to accelerate the exploration of the underlying free-energy landscape within the metadynamics approach. A set of conditional ensembles were collected along reactive paths. The conditional probability distribution function (PDF) of several order parameters was studied for each target system. We will discuss the mechanisms of methane-hydrate nucleation in the heterogeneous environment of silica surfaces and compare these to the homogeneous environment, providing useful insights on nucleation phenomena in both homogeneous and marine environments. This is helpful in the context of elucidating hydrate-nucleation phenomena with a view to potential exploitation of marine-hydrate gas resources.

457287 Molecular Simulations of Wet Flue Gas Adsorption on 13X Zeolite

Abstract:

Molecular Simulations of Wet Flue Gas Adsorption on 13X Zeolite

Mark Purdue,[1,2] Zhiwei Qiao,[1] Jianwen Jiang,[1] Shamsuzzaman Farooq[1,2]

[1] Department of Chemical and Biomolecular Engineering, National University of Singapore, Singapore 117585

[2] Cambridge Centre for Advanced Research in Energy Efficiency in Singapore (CARES), CARES at CREATE Tower. #05-05, 1 CREATE Way, Singapore, 138602

Adsorption technology is one of the leading options currently being explored for carbon dioxide capture and concentration (CCC) from flue gas. Among the adsorption based processes, Vacuum Swing Adsorption (VSA) on 13X Zeolite has been widely studied for CCC from (synthetic) dry flue gas containing 10-15% CO_2 in balance nitrogen. However, real flue gas from any source will always contain moisture. Competition between H_2O and CO_2 for adsorption in 13X Zeolite has a large impact on the VSA process performance, leading to considerations for the use of a guard bed adsorbent such as silica gel or alternative processes to dehydrate the feed gas to a suitable extent. Sacrificing an initial water penetration depth of 13X Zeolite in a single stage VSA process column was recently considered using approximate equilibrium data in VSA optimization studies. Despite the sensitivity of CO_2 capture performance of 13X Zeolite to moisture, wet flue gas on 13X has been addressed only in a few studies using extrapolation of limited available equilibrium data for CO_2/H_2O mixtures. An objective techno-economic evaluation of VSA processes with new materials in varying humidity levels requires a validated simulation model. Process simulation model validation requires availability of sufficient adsorbent material. A process model using a dual-adsorbent VSA process with adsorbents Silica Gel - 13X Zeolite for CCC from wet flue gas can be validated to establish a cost baseline for assessing new materials and process developments. An essential pre-requisite for that is availability of detailed equilibrium data for $CO_2/N_2/H_2O$ mixtures on 13X Zeolite, which was addressed in this study.

Grand Canonical Monte Carlo molecular simulations for the adsorption of CO_2, N_2 and H_2O and associated gas mixtures on 13X Zeolite were performed. The 25Angstrom unit cell of 13X Zeolite adopted contained 83 mobile Na^+ cations, corresponding to a Si/Al ratio of 1.31. Unlike H_2O, CO_2 and N_2 were blocked from entering smaller beta cages using 8 virtual 4Angstrom radius blocking spheres per unit cell, which is illustrated in Figure 1 for N_2. Intermolecular interactions between fixed charge models of adsorbate and adsorbent consisted of 12-6 Lennard-Jones and Coulombic potentials. The isotherms of pure components at 298, 323, and 348 K in the range 0 to 1 atm for CO_2 and N_2, and from 0 to $P^{sat}_{H2O}(T)$ for H_2O were initially simulated and compared to experimental data. Multicomponent adsorption was subsequently investigated at these temperatures and over the full range of compositions to study the influence of water vapour on CO_2 capture equilibrium. This study provides simulated isotherms for comparison with experimental adsorption measurements by both thermogravimetric analysis and dynamic column breakthrough approaches. A mathematical adsorption isotherm model, which characterizes the adsorption of wet flue gas mixtures on 13X Zeolite, is presented to facilitate accurate scale-up of VSA processes for the capture of CO_2 from real flue gas.

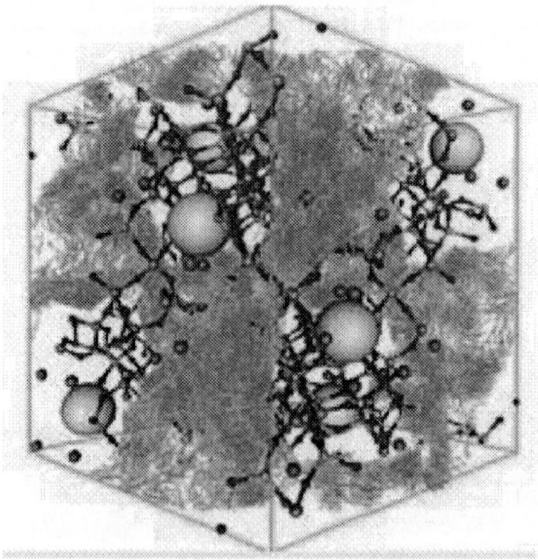

Figure 1: N_2 adsorption (100kPa, 298K) on 13X Zeolite with beta cage blocking

470816 Exploring the Boundaries of Gas Adsorption Via Randomly Generated Porous Materials

Abstract:

There continues to be tremendous interest in studying and developing new porous materials for gas storage and separations applications, and molecular simulations have played an integral role in guiding these efforts. Traditionally, molecular simulations have been used to help understand adsorption behavior in already synthesized materials, but increasingly they are being used to screen *hypothetical* materials *in advance* of their synthesis. Here, we describe a yet new role for molecular simulations: to discover the bounding behavior of gas adsorption in porous materials by screening practically all conceivable materials.

Our approach is to randomly generate a set of porous materials, simulate their gas adsorption behavior, and then generate subsequent populations by mutating those materials whose structure-property combinations were rare. We continue iteratively generating and mutating randomly generated materials until all structure-property combinations are equally represented (at which point the method has converged).

The resulting structure-property data provides a clear picture as to what gas adsorption behavior can practically be discovered in the universe of possible porous materials. We anticipate this approach to have an important impact in determining various â€œbest-caseâ€ scenarios related to important applications such as hydrogen storage, carbon-dioxide capture, and various industrial separation processes.

472928 Interfacial Tension from Simulation and Theory

Abstract:

Of considerable importance in enhanced oil recovery is the interfacial tension between oil and water as well as the rock surface. High interfacial tension between oil and water traps oil droplets in the porous media of reservoirs. Using surfactant formulations, the interfacial tension can be reduced to allow improved oil recovery. As a first step toward modeling these systems, the interfacial tension between model oils and water (specifically toluene and water, dodecane and water, and a mixture of toluene and dodecane with water) are predicted using a series of molecular modeling approaches. In this work we compare predictions based on molecular dynamics simulation, dissipative particle dynamics, density functional theory, and density gradient theory versus laboratory data. The strengths and weaknesses of each of these approaches is discussed.

463563 Gas-Liquid Chromatography Column Selection By Molecular Simulation

Abstract:

Gas-liquid chromatography is widely used for separation based on the variance in free energy of transfer for solutes. In this work, configurational-bias Monte Carlo simulations in the Gibbs ensemble are performed to predict the retention indices of solute molecules. Results can be applied to aid the selection of suitable stationary phases for the separation of gases or alkane isomers using gas-liquid chromatography. Structural analysis can provide additional insight into the mechanism of the separation. Moreover, the retention indices can serve as a stringent validation for the force field used in this study.

469822 Towards the Development of Single Atom Alloy Catalysts As a Means of Escaping Linear Scaling Relations

Abstract:

The rational design of heterogeneous catalysts remains a formidable task, though if successful will pave the way to more efficient and effective catalysts. Linear scaling relations have provided important steps in advancing rational design, significantly facilitating the task faced. For example, the Brønsted-Evans-Polanyi (BEP) relationship correlates the binding energy of a species with its activation energy for a given chemical reaction, allowing for deductions to be made about a variety of catalytic surfaces based on fundamental surface adsorption properties.[1][2] The Sabatier principle connects the BEP relation to catalytic activity yielding Volcano plots which demonstrate that the most active catalysts must bind substrates strongly enough to dissociate reactants but weakly enough to allow for desorption.

Ab initio calculations, using for instance density functional theory (DFT), allow for the simulated study of important catalytic materials. Many such studies on metal surfaces have demonstrated the validity of BEP relationships for numerous simple chemical reactions, allowing for extrapolations and inferences to be made about more complicated, yet related systems. [3-6] The apparent universality of the BEP relation could on one hand be extremely useful in predicting the activity of a catalytic surface, though on the other hand will inevitably limit the prospects of catalyst due to the linearity between activation energy and binding strength.

Many of the catalysts that adhere to BEP type linear models often exhibit spatial homogeneity. Moving away from such surfaces, stepped or kinked surfaces as well as metal alloys have much greater spatial heterogeneity and in many instances still show BEP type trends, for example in the cases of pseudomorphic monolayers and near surface alloys. [7] In this study, we focus on alloys consisting of single, isolated metal dopant atoms dispersed in the surface layer of a host metal; these materials are referred to as single atom alloys (SAAs). High single atom dispersity in the surface of a SAA leads to unique strain and ligand effects unseen in pure metal analogues and many other alloy structures, giving rise to some counterintuitive surface binding properties. In addition, the high degree of spatial heterogeneity in the surface of a SAA may limit the description of its behaviour by simple BEP relations. It has been shown in several systems that SAAs can adsorb and dissociate reactants at one site before facile spillover to the host metal allows for further reaction to occur [8-11].

We present a DFT-based study of nine SAAs formed from Au, Ag and Cu hosts doped with individual surface atoms of Ni, Pd and Pt. We investigate the chemistries of catalytically relevant bond dissociation reactions such as in the scission of C-H, N-H, H-H, O-H, C-C, N-N and O-O. Our results illustrate how strain and ligand effects contribute to the reactivity of the dopant atom, showing that different host metals can result in reduced or increased reactivity compared to the pure dopant metal surfaces. We demonstrate that isolated single atoms are sufficient for such chemistries and also allow for facile spillover onto host metal facets. Moreover, we rigorously evaluate the validity of the BEP relationship on each SAA for each process, determining the cause of any alloys not following a simple linear correlation.

References

1. Evans, M. G., Polanyi, M., *Trans. Faraday Soc.,* 1938, **34**, 11.
2. Brønsted, N., *Chem. Rev.*, 1928, **5**, 231.
3. Nørskov, J .K., Bligaard, T., Logadottir, A., Bahn, R. S., Hansen, L. B., Bollinger, M., Bengaard, S. H., Hammer, B., Slijivancanin, Z., Mavrikakis, M., Xu, Y., Dahl, S., Jacobsen, C. J. H., *J. Catal.*, 2002, **209**, 275.
4. Logadottir, A., Rod, T. H., Nørskov, J .K., Hammer, B., Dahl, S., Jacobsen, C. J. H., *J. Catal.*, 2001, **197**, 229.
5. Michaelides, A., Liu, Z. P., Zhang, C. J., Alavi, A., King, D. A., Hu, P., *J. Am. Chem. Soc.*, 2003, **125**, 3704.
6. Wang, S. Temel, B., Shen, J., Jones, G., Grabow, L., Studt, F., Bligaard, T., Ablid-Pedersen, F., Christensen, C. H., Nørskov, J .K., *Catal. Lett.*, 2010, **141**, 370.
7. Nørskov, J .K., Bligaard, T., Rossmeisl, J., Christensen C.H., *Nat. Chem.*, 2009, **1**, 37.
8. Baber, A. E., Tierney, H. L., Lawton, T. J. and Sykes, E. C. H., *Chem. Cat. Chem.*, 2011, **3**, 607.
9. Tierney, H., Baber, A., Kitchin, J., Sykes, E. C., *Phys. Rev. Lett.*, 2009, **103**, 246102.
10. Marcinkowski, M. D., Jewell, A. D., Stamatakis, M., Boucher, M. B., Lewis, E. A., Murphy C. J., Kyriakou, G., Sykes, E. C. H., *Nat. Mater.*, 2013, **12**, 523.
11. Lucci, F., Darby, M., Mattera, M., Ivimey, C., Therrien, A., Michaelides, A., Stamatakis, M., Sykes, E. C., *J. Phys. Chem. Lett.*, 2016, **7**, 480.

464611 Theory of Split Quantum Dot Formation in Strained-Layer Semiconductor Heteroepitaxy

Abstract:

Semiconductor quantum dots (QDs) are three-dimensional (3D) nanocrystalline structures with a broad range of applications in the fabrication of electronic, optoelectronic, and photovoltaic devices. Controlling the size of quantum dots is used to tailor their electronic structure and has direct impact on their optoelectronic properties. A common approach to QD growth is strained-layer heteroepitaxy through the Stranski-Krastanow (SK) instability of the epitaxial film surface morphology, where the competition between surface energy and elastic strain energy in the epitaxial film results in a growth transition, from layer-by-layer epitaxy to formation of 3D islands. Intriguing, complex QD morphologies have been observed experimentally, such as those known in the literature as "double quantum dots" or "quantum dot pairs" characterized by split tips of variable height. This interesting pattern forming phenomenon can be exploited to control the size of QDs grown by strained-layer heteroepitaxy; however, such QD splitting is not well understood.

Toward a fundamental understanding of the quantum dot tip-splitting phenomenon, in this presentation, we report the results of a weakly nonlinear surface morphological stability analysis of a coherently strained heteroepitaxial thin film on a crystalline substrate. The analysis is based on a fully nonlinear continuum model of film surface morphological evolution coupled with the elastostatic boundary-value problem to determine the stress state in the epitaxial film. The model also accounts for a wetting potential contribution to the free energy of the film as well as surface diffusional anisotropy. The weakly nonlinear theory predicts that, in addition to the SK instability, long-wavelength perturbations from the planar film surface morphology also can trigger nonlinear instabilities, resulting in the splitting of a single QD into multiple QDs of smaller sizes. The theory also predicts the critical wavelength of the film surface perturbation for the onset of the nonlinear tip-splitting instability. We have found that, although the tip-splitting instability is caused entirely by the competition between the surface energy and the elastic strain energy, the wetting potential has a stabilizing effect on the planar film surface morphology, increasing the critical wavelength for the onset of the nonlinear tip-splitting instability. This critical wavelength is not affected by surface diffusional anisotropy. Moreover, we have conducted a systematic protocol of self-consistent dynamical simulations of the epitaxial film surface morphological evolution to assess the validity of the theoretical results. We have simulated QD tip splitting as a perturbed film planar surface evolves morphologically and validated the predictions of the weakly nonlinear stability theory. We have found that surface diffusional anisotropy affects the morphology of the QD, resulting in an asymmetric splitting of the QD and leading to QD morphologies that resemble strongly the TEM and AFM images of double QDs or QD pairs reported in the literature. Our study sets the stage for exploiting nonlinear surface phenomena in strained-layer heteroepitaxy toward precise engineering of tunable-size nanoscale surface features for optimal optoelectronic function.

468139 The Effect of Select Organic Compounds on Hydration of Portland Cement: Experimental and Molecular Dynamics Study

Abstract:

A large sector of industrial chemicals are amphiphilic organic compounds. When dissolved in water, amphiphlies are attracted to the air-water, i.e. nonpolar-polar, interface. In the concrete industry, amphiphilic compounds are used to reduce shrinkage cracking in portland cement concrete by reducing the surface tension of the pore water within the cement microstructure. The concrete pore solution contains sodium, potassium, calcium and to a lesser extent other ions, largely due to the chemistry of portland cement. Assembly and concentration of the amphiphilic compound at the air-water interface in the presence of various ions is critical to the understanding of compound effectiveness. In addition to that, amphiphilic compounds are not chemically or physically inert and can negatively interfere with the progress of cement hydration and at the same time be effective surface tension reducing agents. Side effects such as hydration retardation lead to delayed strength development or worse. The influence of the amphiphilic compounds on the cement product phases is a complex multi-parameter problem.

To understand the effect of organic compounds on hydration of cement, a combined experimental and computational approach was used. This study summarizes ongoing simulations and experiments using three compounds: butyl glycolate, 2-butyoxyethyl acetate, and hexylene glycol. Molecular dynamics simulations were performed to simulate the interactions of the test compounds with the (001) surface of the portlandite crystal (calcium hydroxide) and (040) surface of the tricalcium silicate crystal (the most abundant species in ordinary portland cement). In addition, further molecular dynamics simulations were performed to investigate the interactions of the test compounds with simulated pore solution. To measure the adsorption isotherm of chelating agents - calcium glycolate and calcium acetate - experiments were performed by equilibrating aqueous solutions of the agents in the presence of various amounts of solid phase calcium hydroxide. In addition, isothermal calorimetry, gas chromatography and surface tension experiments were also used to elucidate the process and support the computational work.

452689 Accurate Property Prediction for Inorganic Materials with Machine Learning

Abstract:

Historically, novel materials have been discovered because of long and laborious trial-and-error process. Over the years, materials research has led to the accumulation of relatively large collections of experimental data on materials structure and properties. Prompted by the growth of materials databases, the emerging materials informatics approaches offer an opportunity to transform this trial-and-error practice into data- and knowledge-driven rational design and accelerated discovery of novel materials with the desired properties.

Using the data from the AFLOWLIB repository (http:www.aflowlib.org) of materials properties obtained with the high-throughput DFT calculations, we have constructed Machine Learning (ML) models to predict three critical materials properties including band gap, Fermi level energy, and the class of materials as metals or insulators. To enable these calculation, we have developed novel materials descriptors such as universal property-labelled fragments (PLMF).[1] We have established that the accuracy of predictions obtained with Quantitative Materials Structure–Property Relationship (QMSPR) models approaches that of GGA DFT functionals yet model development requires a minute fraction of computational time as compared to ab initio calculations. Notably, due to the representation of materials with PLMF the QMSPR models are broadly applicable to virtually any stoichiometric inorganic materials. This representation also affords straightforward model interpretation in terms of simple heuristic design rules that could guide rational design of novel materials. This proof-of-concept study demonstrates the power of materials informatics to dramatically accelerate the search for novel materials.

[1] O. Isayev, C. Oses, S. Curtarolo, A. Tropsha. The materials genome of electronic structure: prediction of band gap with machine learning. *In press*

457255 Comparing Molecules and Solids Across Structural and Alchemical Space

Abstract:

Evaluating the (dis)similarity of crystalline, disordered and molecular compounds is a critical step in the development of algorithms to classify structures, search chemical space for better compounds and materials, and drive the next generation of machine-learning algorithms for predicting the stability and properties of atomic systems. In recent years several strategies have been designed [1-3] to compare atomic coordination environments. In particular, the Smooth Overlap of Atomic Positions has emerged as a natural framework to obtain translation, rotation and permutation-invariant descriptors of atomic environments, driven by the design of various classes of machine-learned inter-atomic potentials. Here we will present few examples showcasing how one can construct a Sketchmap[4-6] representation of databases of both molecular and bulk structures, using (dis)similarity definitions based on such local descriptors that can treat alchemical and structural complexity within a unified framework.

[1] A. P. Bart ok, et al, Phys. Rev. B88, 054104(2013) [2] Ali Sadeghi et al, J. Chem. Phys. 139, 184118 (2013) [3] Sandip De et al, Phys. Rev. Lett. 112, 083401(2014) [4] G. A. Tribello et al,Proc. Acad. Natl. Sci. U.S.A. 109 5196 (2012) [5] M. Ceriotti et al, Proc. Acad. Natl. Sci. U.S.A. 108 13023 (2011) [6] M. Ceriotti et al, J. Chem. Theory Comput. 9 1521 (2013)

461815 Discriminative Neural Embeddings of Latent Variable Models for Molecular Property Prediction

Abstract:

Kernel classifiers and regressors designed for structured data, such as sequences, trees and graphs, have significantly advanced in a number of interdisciplinary areas such as computational biology and dug design. Typically, the kernel functions are designed beforehand for a data type which either exploit statistics of the structures or make use of probabilistic generative models, and then a discriminative classifier is learned based on the kernels via convex optimization. However, such an elegant two-stage approach also limited kernel methods in terms of their ability to scale up to millions of data points and exploit discriminative information to learn feature representations.

We propose an effective and scalable approach for structured data representation which is based on the idea of embedding latent variable models into feature spaces, and learning such feature spaces with ultimate regressor/classifier using discriminative information. The algorithm runs a sequence of function mappings in a way similar to graphical model inference procedures, such as mean field and belief propagation. This can be implemented as a recurrent neural network, where the parameters for the feature spaces and classifiers are trained in an end to end fashion. We deployed our algorithm on several computational chemistry problems, including the compound and protein classification tasks. We achieved state-of-the-art results on several commonly used benchmark datasets, including NCI1, NCI90, ENZYMES and D&D. We also applied our algorithm on Harvard Clean Energy Project dataset with millions of molecules and predicted power conversion efficiency and energy. We achieved below 0.1 mean absolute error, while using significantly small number of parameters than alternatives.

470152 On-the-Fly Heuristic Reordering Approach to Deterministic Optimization for Qualitative Chemical Property Prediction

Abstract:

Chemical optimization and design affords current and future researchers the ability to harness the potential of chemical space towards the fast and efficient discovery of novel materials. In the present work, a multi-constraint deterministic optimization technique based on the on-the-fly heuristic reordering of chemical subspace has been developed and used towards materials discovery in several systems of interest. The competitive advantage of the deterministic optimization method results from the combination of fast computational techniques and innovative design algorithms which allow for intelligent screening of a large number of chemical compounds within a reasonable computational time. A family of search algorithms has been used to approach the problem of navigating chemical subspace, including general base line search and general base gradient local search techniques. Because ideal ordering of the chemical subspace is not known, measures must be taken to ensure that the space is being sampled properly. There are several strategies utilized in this work to ensure this requirement is being satisfied. First, optimization algorithms based on heuristic reordering of the chemical subspace are developed to assist and direct the optimization procedure. The heuristic reordering algorithms play an essential role in optimization efficiency and each heuristic scheme has been uniquely developed as an effort to further enhance the subspace sampling. Each of these algorithms have been benchmarked and tested for their performance with respect to candidate structure discovery. In addition, to further combat the potential subspace sampling partiality, a binary entropic, enhanced sampling approach has been employed. This technique allows for generation and searching of structures which are chemically maximally different from the local best candidate structure. This is advantageous because a larger breadth of space is able to be sampled in an equally efficient manner. This method has been applied to several systems of interest including high-hyperpolarizability materials, energetic materials and optically switchable materials. Detailed analysis was performed for each of these systems and qualitative structure property relationships were determined.

461054 Machine Learning for Predicting Accurate Quantum Chemical Energies

Abstract:

Reliable *in silico*drug and materials design requires accurate yet fast methods for high-throughput screening. The most accurate high-level quantum chemical methods are generally too slow for that purpose. We demonstrated that accurate quantum chemical energies can be predicted by combining fast machine learning (ML) and low-level quantum mechanics (QM) methods given sufficiently large training sets.

One of such hybrid QM/ML techniques directly corrects energies calculated with low-level QM methods using ML (Î”-ML approach). Fast density functional theory (DFT) or semiempirical quantum chemical (SQC) methods can be used as low-level QM methods. Î”-ML approach readily achieves chemical accuracy (error ca. 1Â kcal/mol).[1]

Another approach is to improve semiempirical Hamiltonian itself by using ML to predict *on-the-fly* SQC parameters for individual molecules (automatic parametrization technique, APT). APT stands in stark contrast to the traditional special-purpose reparametrization (SPR), where parameters are optimized for specific type of molecules and used unchanged for every other target molecule. Thus APT has several advantages over SPR: its accuracy can be further improved by increasing training set size, it is essentially black boxed, molecules far outside the training set are calculated with accuracy of SQC method with standard parameters.[2]

Finally, improved ML-based techniques can be used for very fast and accurate modeling of molecular potential energy surfaces, which are used to calculate highly accurate rovibrational spectra of small molecules (error ca. 1Â cm$^{\text{â}''1}$).[3]

[1] R. Ramakrishnan, P. O. Dral, M. Rupp, O. A. von Lilienfeld, *J. Chem. Theory Comput.* **2015**, *11*, 2087.

[2] P. O. Dral, O. A. von Lilienfeld, W. Thiel, *J. Chem. Theory Comput.* **2015**, *11*, 2120.

[3] P. O. Dral, A. Owens, S. N. Yurchenko, W. Thiel, *in preparation*.

452723 Deductive and Inductive Modeling of Electronic Properties in the Organic Molecular Design Space

Abstract:

Socio-economically important researches such as designing single-molecular electronic devices, depend on reliable virtual screening of electronic properties in the chemical space of organic molecules. Clearly, for this task, the only viable solution is to employ statistical property models trained on quantum chemistry datasets. For a given modeling framework, e.g. kernel-ridge-regression with a specified kernel function, the most efficient strategy is to "delta-learn", i.e., to train the model on the prediction errors of fast, but qualitatively accurate, legacy quantum chemistry methods -- rather than directly on outcomes from accurate many-body methods. I will present proof-of-concepts, using non-trivial "Big Datasets", for accurate "delta-learning" of several molecular properties, with hierarchially improved quantum chemistry methods. Latest results on novel electronic properties of isolated molecules and their composite forms will also be discussed.

References

1. Ramakrishnan et al., "Quantum Chemistry Structures and Properties of 134 Kilo Molecules", Scientific Data 1 (2014).
2. Ramakrishnan et al., "Big Data Meets Quantum Chemistry Approximations: The Delta-Machine Learning Approach", Journal of Chemical Theory and Computation 11 (2015) 2087-2096.
3. Ramakrishnan et al., "Many Molecular Properties from One Kernel in Chemical Space", Chimia International Journal for Chemistry 69 (2015) 182-186.
4. Ramakrishnan et al., "Electronic Spectra from TDDFT and Machine Learning in Chemical Space", The Journal of Chemical Physics 143 (2015) 084111.
5. Ramakrishnan et al., "Machine Learning, Quantum Mechanics, and Chemical Compound Space",

   arxiv:1510.07512.

462108 Sorting out a Process-Structure-Property Relationship in Polymer Organic Electronics

Abstract:

The world is full of data, but automated extraction of knowledge from this data is often difficult. In this talk the focus is on processing of polymer organic electronic devices. Two challenges that will be discussed are: how to extract meaningful and quantitative structure-property relationships from high-dimensional AFM data and how to learn process-property relationships from a literature database. Both challenges will be discussed using poly-3-hexylthiophene (P3HT) as a case study.

A semi-automated approach to AFM structure quantification will be presented. This method enabled the assignment of process-property variability in spin-coated P3HT organic field effect transistors (OFETs) to the process-structure relationship. More importantly, the structure-property relationship was elucidated and had less variability. Meso-scale orientational order of P3HT nanofibers was shown to be a quantitative predictor of hole mobility in P3HT OFETs, arising from a combination of solution aging and the spin-coating flow field.

A tremendous amount of experimental literature exists on P3HT processing. However, many challenges remain in comparing "apples to apples." The property of interest, hole mobility, is influenced by starting material properties, solution preparation, deposition method, and environmental factors during characterization. These experimental features are inconsistently reported by researchers, creating a sparse matrix of process-property data in the literature. A sortable database was created to peruse the literature and compare devices across studies.

Ultimately we determined that the best use of the literature database may be to generate new hypotheses for additional experimental studies. These hypotheses can be encoded in mathematical models, which are then used in an experimental design algorithm to calculate the "best" next experiments to perform. Once the data is obtained, it can be used to update the models, and repeat the cycle. Tailored methods for sequential statistical design will be presented here to formalize this linkage between models and experiments.

470968 Application of High Performance Computing and Machine Learning to Accelerate Material Discovery for Energy Capture and Storage

Abstract:

High performance computing (HPC) and open-source software are revolutionizing the strategies for material discovery. In contrast to material discovery in the lab, which involves at every stepâ€"design, implementation, and testingâ€"the ardent attention of the scientist, material discovery driven by the best use of digital resources shifts the focus of the researcher to the creation of automated and adaptive experimental architectures. Such architectures built on the foundations of machine learning (ML), HPC, and database management and allow the exploration of design spaces that would otherwise be insurmountable.[1]

The design space for materials in alternative energy is one such space that is simply too vast to be explored by laboratory methods alone. The first part of this talk will discuss ways that ML algorithms can be applied to the experimental characterization of new materials like solar cells. In particular, we employ the random forest regressor in the scikit-learn python package to predict the photoluminescence properties of perovskites. The random forest is trained on experimentally measured photoluminescence, height, potential, phase, and amplitude atomic force microscopy (AFM) images and then predicts the photoluminescence using the latter four AFM images as inputs to the learning algorithm. Random forests are convenient for simple applications such as this as they require little parameterization from the user; are easy to interpret as they can be visualised; require comparatively little data preparation as most learning algorithms need input data to be preprocessed in some form; scale logarithmically with the number of data points the tree is trained on; and is a white box model such that if a pattern is to be found among the data it can be articulated by the boolean logic generated by the tree (vs black box models such as artificial neural networks where relationships between features of the data are highly convoluted).

Following this, we will describe how we are applying data science tools to the design of new electrolyte solutions for batteries. Our entire discovery process takes place within HPC: we design an all-atom electrolyte solution, perform molecular dynamics (MD) simulations to measure the physical properties that are of interest, and then perform a search algorithm to increment in a single dimension of the solution before reiterating. In this case, we search for the eutectic ratio of an ionic liquid (IL) to a hydrogen bond donor. These so-called deep eutectic solvents (DESs) have recently been shown to be promising charge-carrying fluids for redox flow batteries but require further reduction in viscosity before they will be viable commercial products. While this one dimensional search allows us to find eutectic ratios for a given IL and hydrogen bond donor, it does not address the much larger design space of the individual components themselves. Extensions of our approach to larger design spaces via the use of artificial neural networks (ANN) will also be discussed. The talk will end with a short perspective on how these tools can be further generalized and applied to a wide range of problems in Chemical Engineering.

[1] Beck, D. A. C., Carothers, J. M., Subramanian, V. R. and Pfaendtner, J. (2016), Data science: Accelerating innovation and discovery in chemical engineering. AIChE J., 62: 1402â€“1416. doi:10.1002/aic.15192

460962 A Combinatorial Approach for Developing Minimally-Parameterized and Highly-Transferable Density Functionals

Abstract:

Conventionally, the development of semi-empirical density functionals has involved selecting a handful of functional forms and optimizing the parameters via least-squares fits to accurate data. However, such a procedure does not ensure that the resulting functional will be transferable to systems outside of its training set. In addition, overfitting to training data can lead to both poor performance on independent test cases and unwanted unphysical characteristics.

Since semi-empirical density functionals utilize power series enhancement factors, the systematic optimization of a density functional can be imagined as a combinatorial search problem, with all possible combinations of the available power series variables serving as potential functional forms. This approach increases the number of candidate functional forms from just a handful to 2^n-1, where n is the total number of available variables. With the significantly enlarged functional space, fits can be characterized not only based on their training set performance, but additionally with respect to their transferability to an independent test set as well as their physical characteristics.

This procedure has recently been used to partially explore a meta-GGA functional space of 10^{41} possible functional forms, resulting in the development of ωB97M-V, a range-separated hybrid, meta-GGA density functional with VV10 nonlocal correlation. With only 12 linear parameters, the functional significantly outperforms methods in and below its class that have upwards of 50 linear parameters on non-covalent interactions, isomerization energies, thermochemistry, and barrier heights. The merits of ωB97M-V will be demonstrated by comparing it to leading density functionals (such as M06-2X and M11) across a database of nearly 5,000 data points.

Efficient *K*-Point Grid Generation through the Use of Informatics

Pandu Wisesa, Kyle McGill, and Tim Mueller
Department of Materials Science and Engineering
Johns Hopkins University

Funded by NSF DMR-1352373

K-Points

JOHNS HOPKINS
WHITING SCHOOL of ENGINEERING

Reciprocal space **Real space**

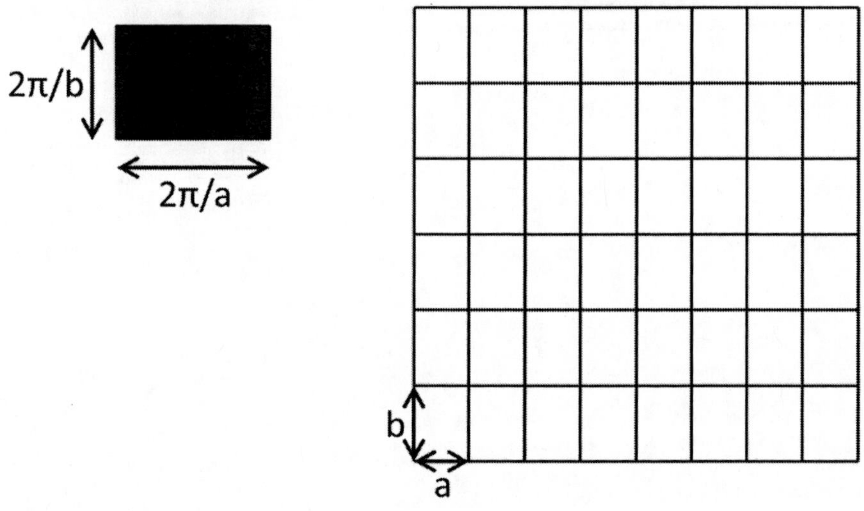

K-Points

Reciprocal space

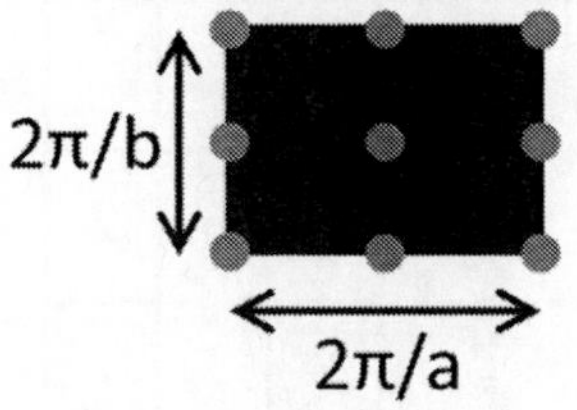

Real space

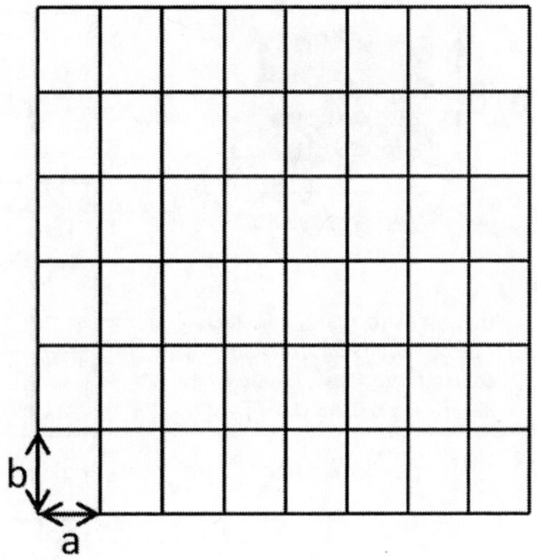

K-Points

Reciprocal space

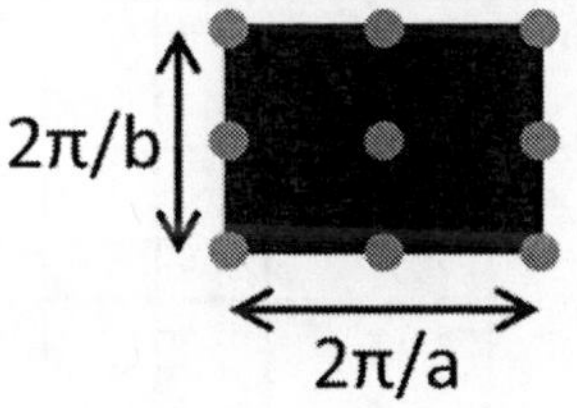

Real space

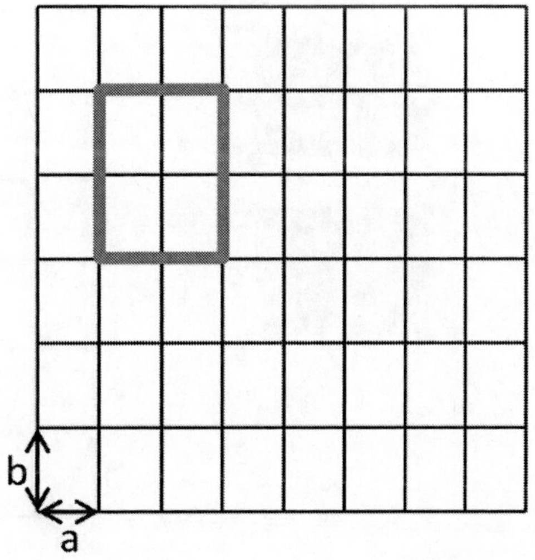

K-Points

JOHNS HOPKINS
WHITING SCHOOL *of* ENGINEERING

Reciprocal space

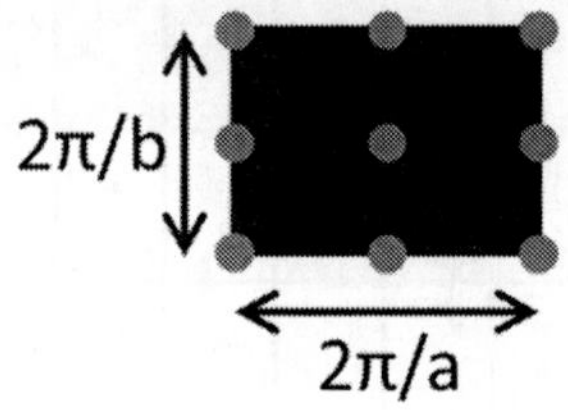

H. J. Monkhorst and J. D. Pack, *"Special Points for Brillouin-Zone Integrations"*, Physical Review B **13**, 5188 (1976).

Real space

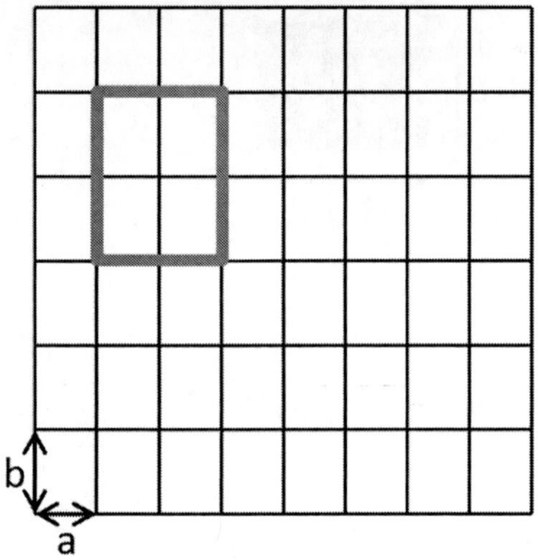

K-Points – Monkhorst-Pack Grids

JOHNS HOPKINS
WHITING SCHOOL *of* ENGINEERING

Reciprocal space

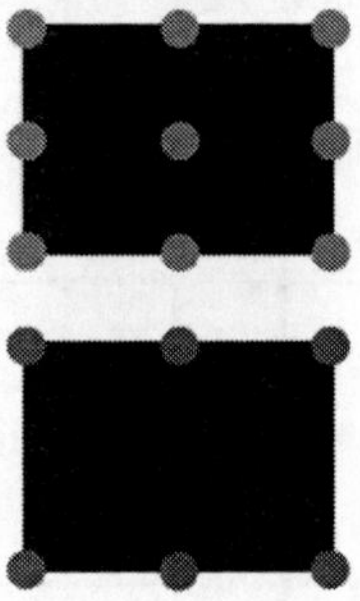

Real space

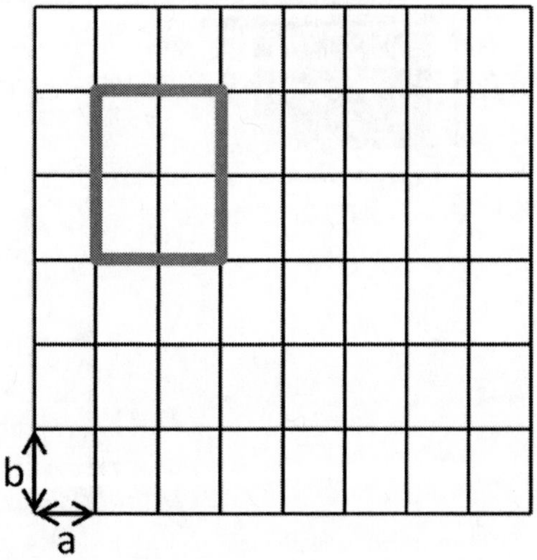

K-Points – Monkhorst-Pack Grids

JOHNS HOPKINS
WHITING SCHOOL of ENGINEERING

Reciprocal space | **Real space**

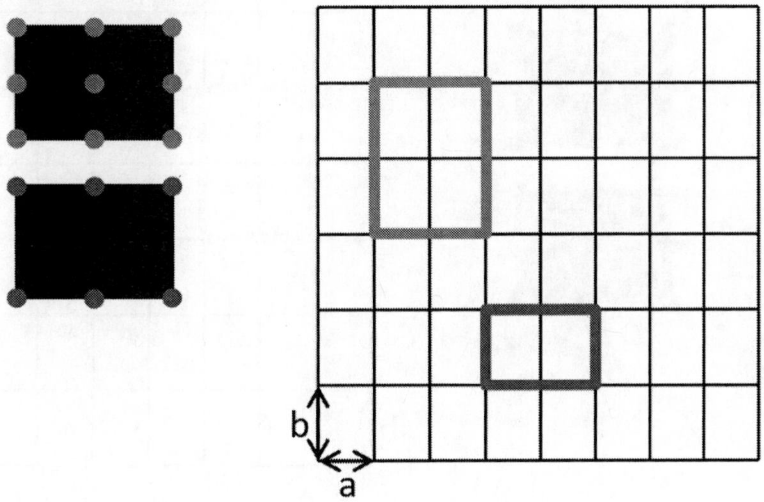

K-Points – Monkhorst-Pack Grids

JOHNS HOPKINS
WHITING SCHOOL of ENGINEERING

Reciprocal space | **Real space**

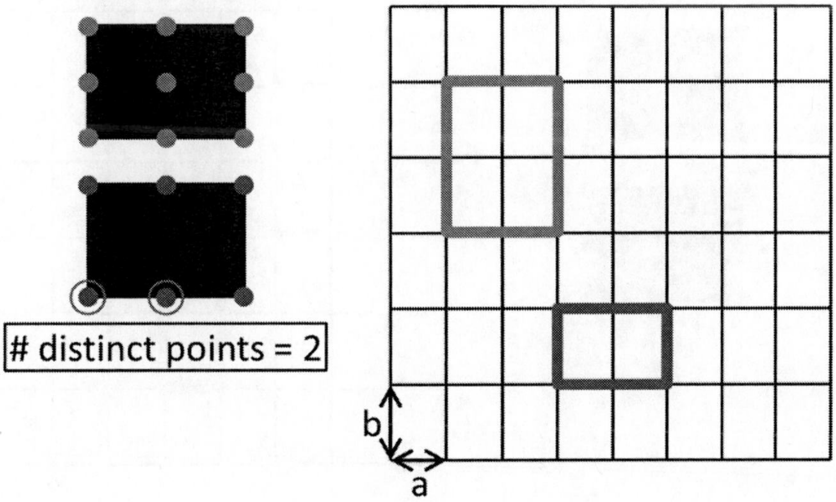

K-Points – Monkhorst-Pack Grids

JOHNS HOPKINS
WHITING SCHOOL of ENGINEERING

Reciprocal space

Real space

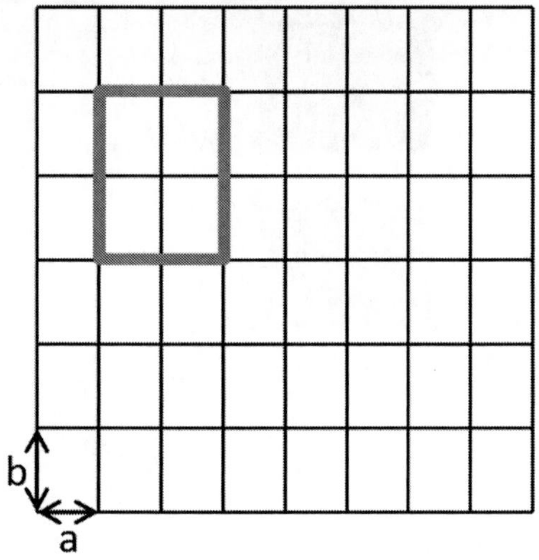

K-Points – Monkhorst-Pack Grids

JOHNS HOPKINS
WHITING SCHOOL of ENGINEERING

Reciprocal space

Real space

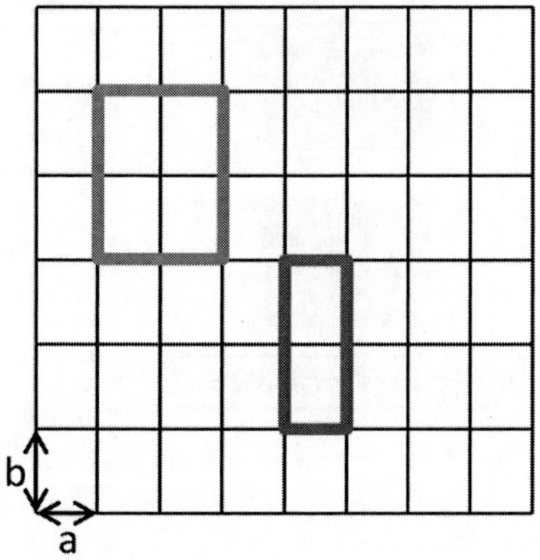

K-Points – Monkhorst-Pack Grids

JOHNS HOPKINS WHITING SCHOOL of ENGINEERING

Reciprocal space **Real space**

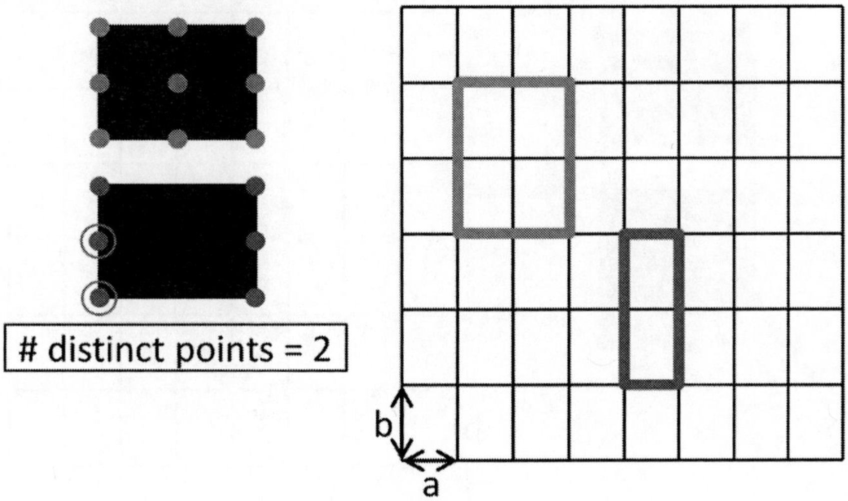

K-Points – Generalized Monkhorst-Pack Grids

JOHNS HOPKINS WHITING SCHOOL of ENGINEERING

Reciprocal space **Real space**

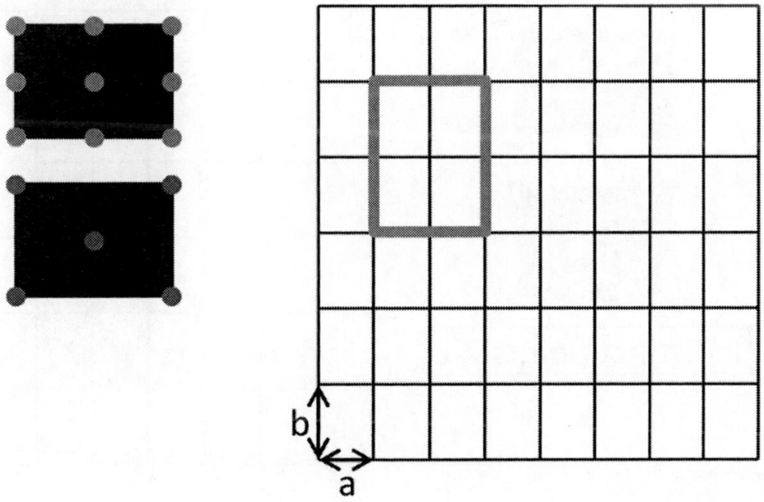

K-Points – Generalized Monkhorst-Pack Grids

JOHNS HOPKINS WHITING SCHOOL *of* ENGINEERING

Reciprocal space

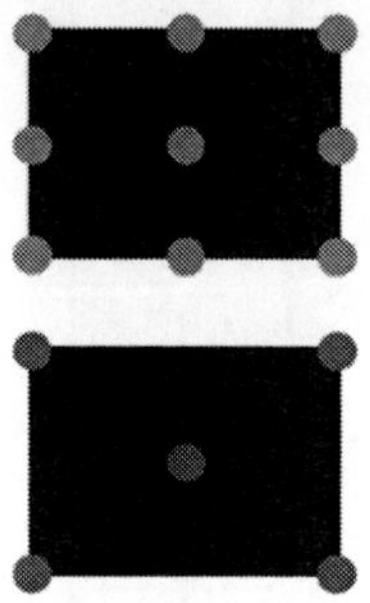

Real space

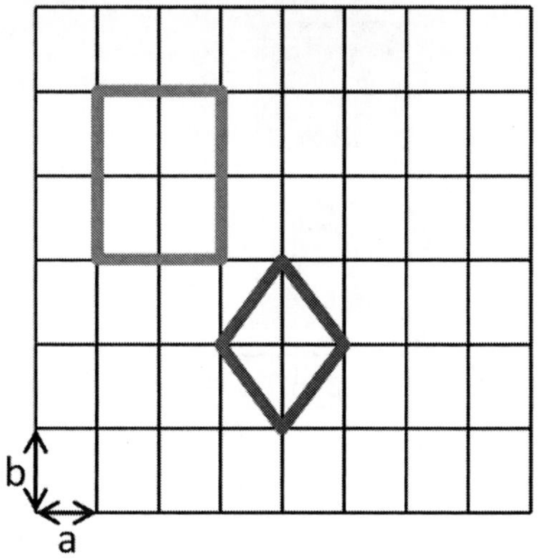

K-Points – Generalized Monkhorst-Pack Grids

JOHNS HOPKINS WHITING SCHOOL *of* ENGINEERING

Reciprocal space

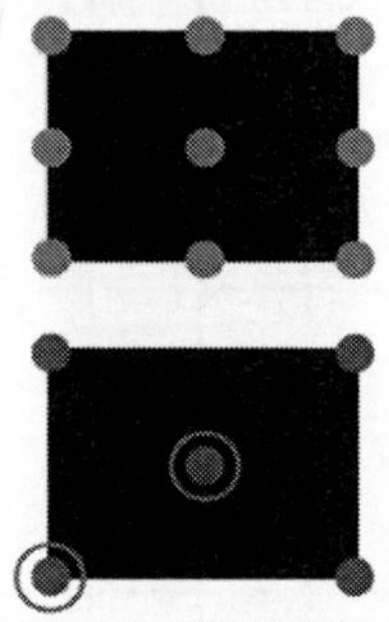

distinct points = 2

Real space

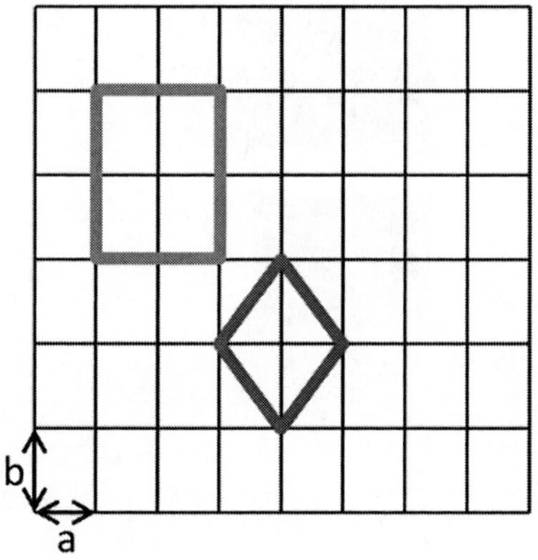

K-Points – Generalized Monkhorst-Pack Grids

JOHNS HOPKINS
WHITING SCHOOL of ENGINEERING

Reciprocal space **Real space**

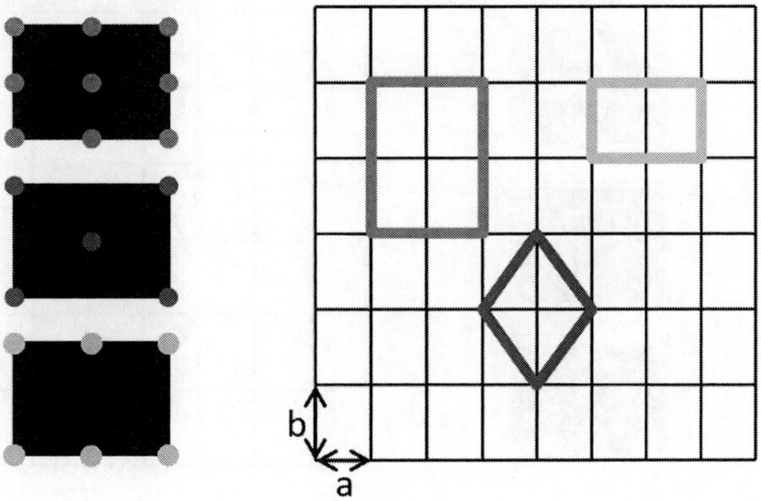

K-Points – Generalized Monkhorst-Pack Grids

JOHNS HOPKINS
WHITING SCHOOL of ENGINEERING

Reciprocal space **Real space**

Same density better sampling.

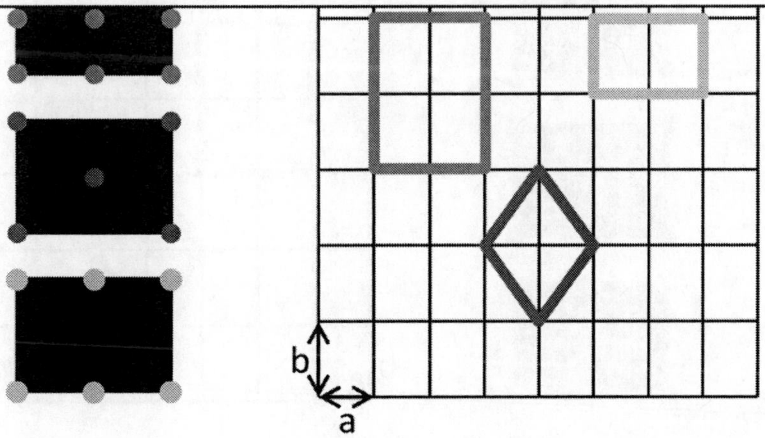

K-Points – Quantifying Even Sampling

Johns Hopkins
Whiting School of Engineering

Reciprocal space **Real space**

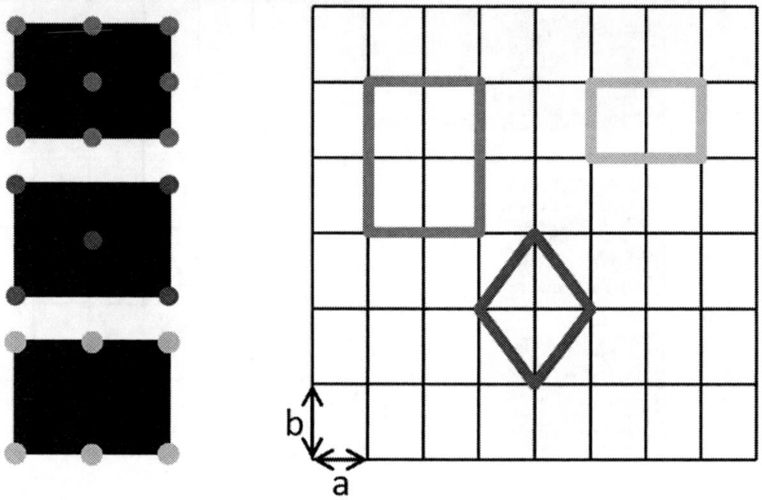

K-Points – Quantifying Even Sampling

Johns Hopkins
Whiting School of Engineering

Reciprocal space **Real space**

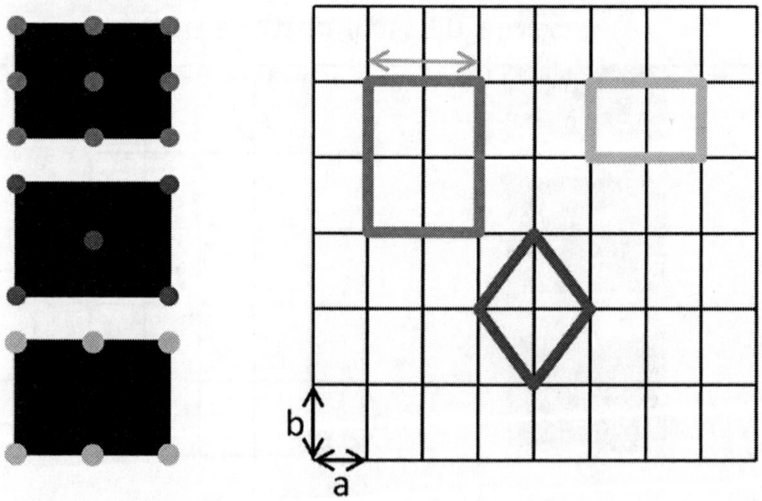

K-Points – Quantifying Even Sampling

JOHNS HOPKINS
WHITING SCHOOL of ENGINEERING

Reciprocal space **Real space**

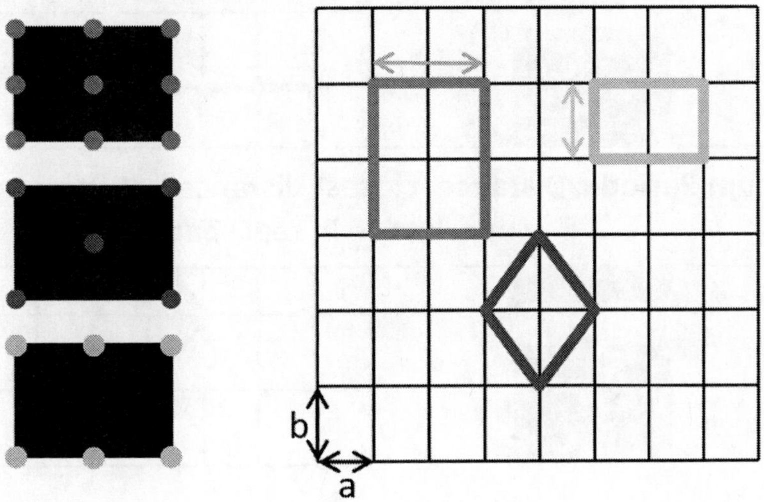

K-Points – Quantifying Even Sampling

JOHNS HOPKINS
WHITING SCHOOL of ENGINEERING

Reciprocal space **Real space**

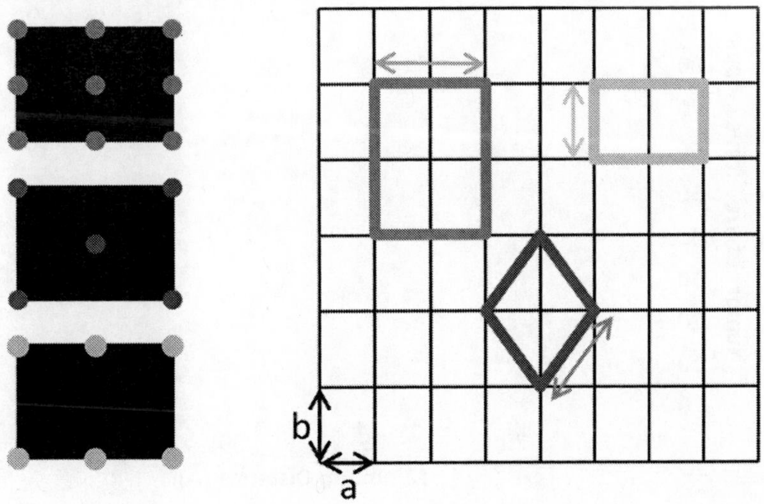

K-Points – Quantifying Even Sampling

JOHNS HOPKINS
WHITING SCHOOL of ENGINEERING

Reciprocal space **Real space**

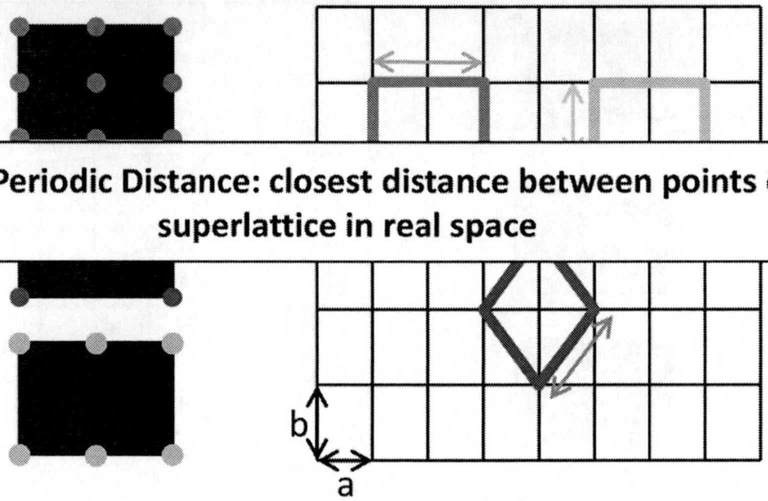

Minimum Periodic Distance: closest distance between points on the superlattice in real space

Combining Point Density and Minimum Distance

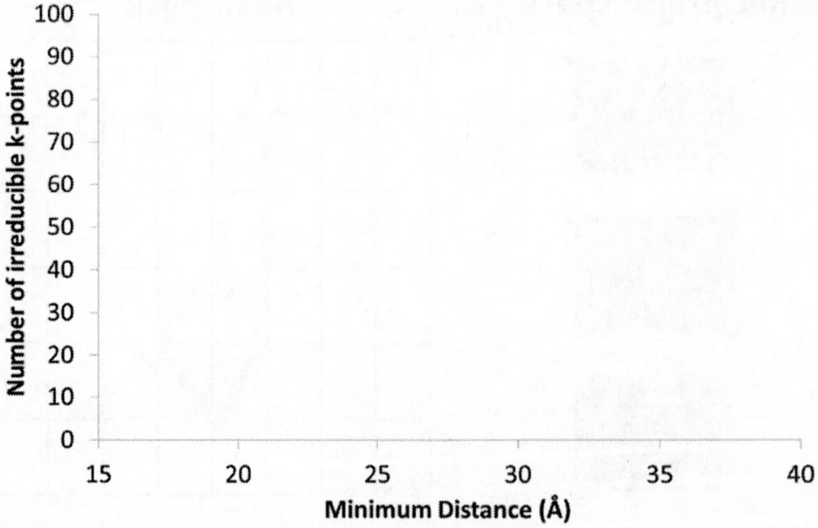

Combining Point Density and Minimum Distance

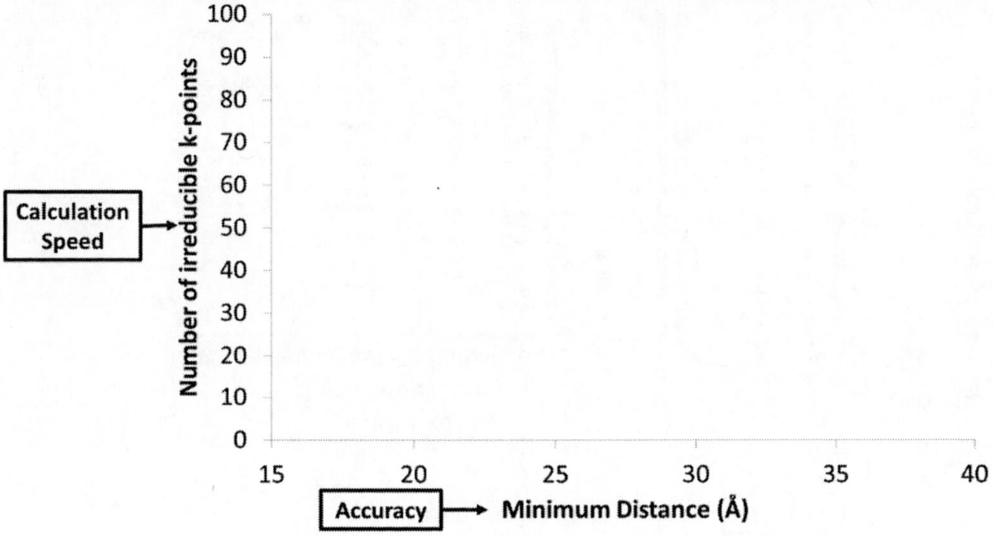

Listing All Possible Lattices

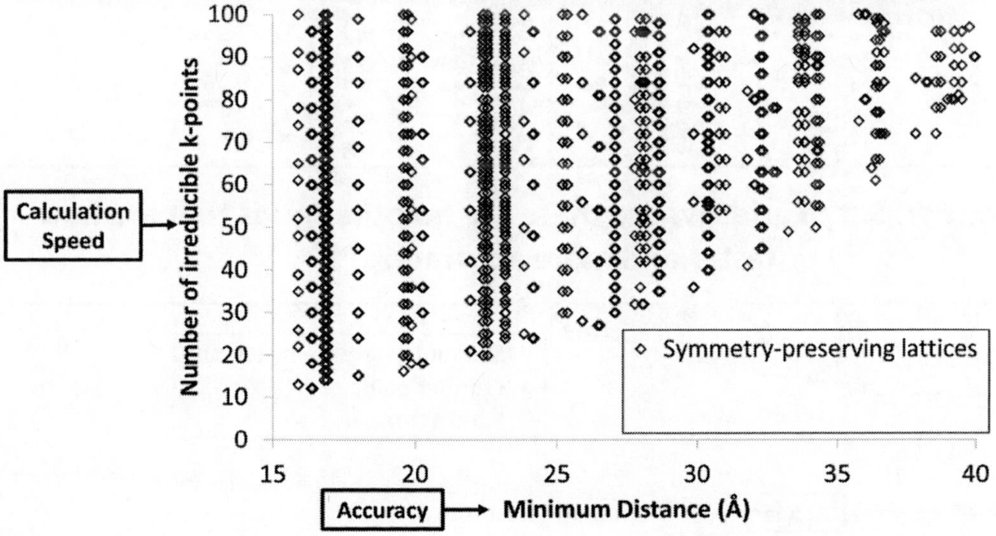

Pareto Frontier

JOHNS HOPKINS
WHITING SCHOOL of ENGINEERING

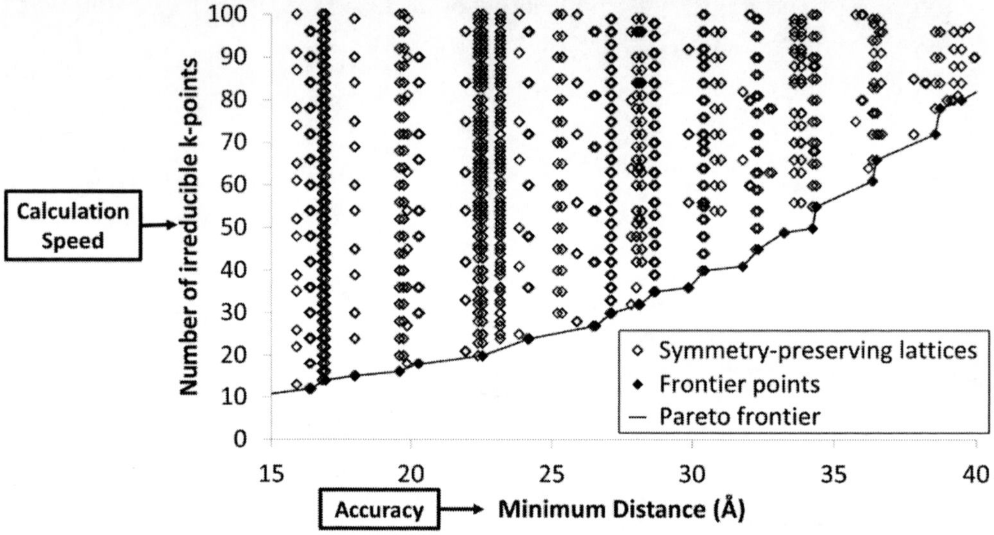

Pareto Frontier

JOHNS HOPKINS
WHITING SCHOOL of ENGINEERING

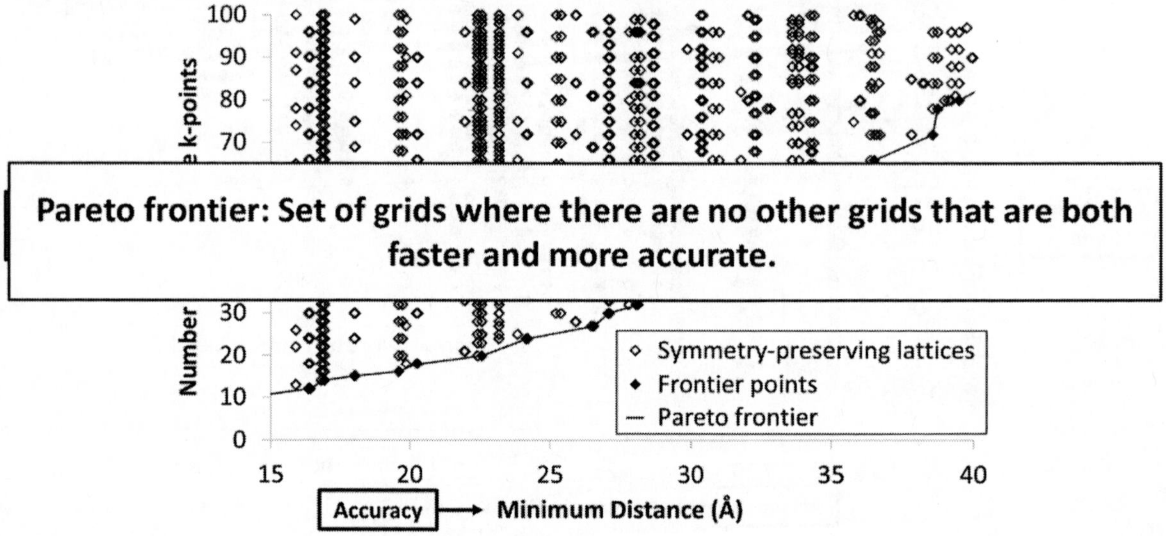

Pareto frontier: Set of grids where there are no other grids that are both faster and more accurate.

Pareto Frontier

JOHNS HOPKINS WHITING SCHOOL of ENGINEERING

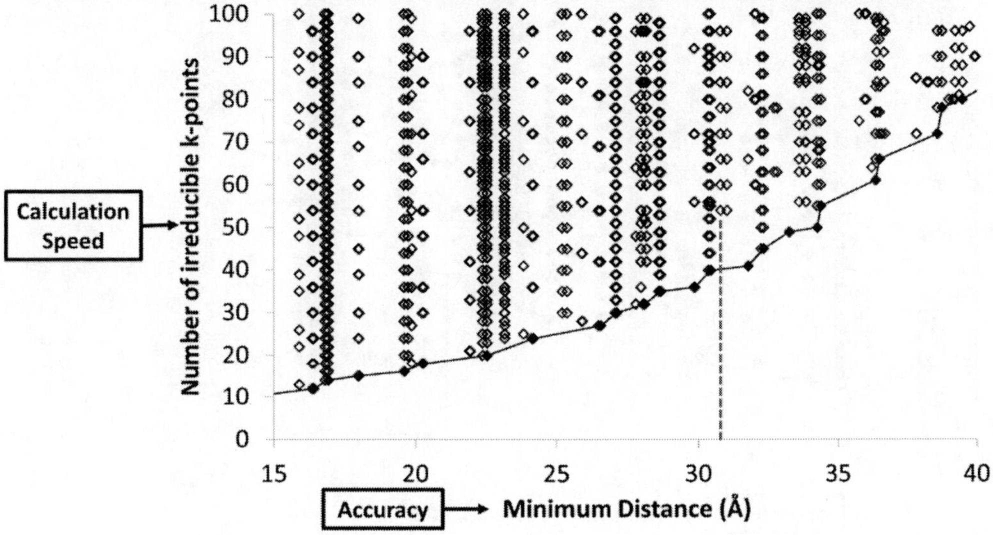

Pareto Frontier

JOHNS HOPKINS WHITING SCHOOL of ENGINEERING

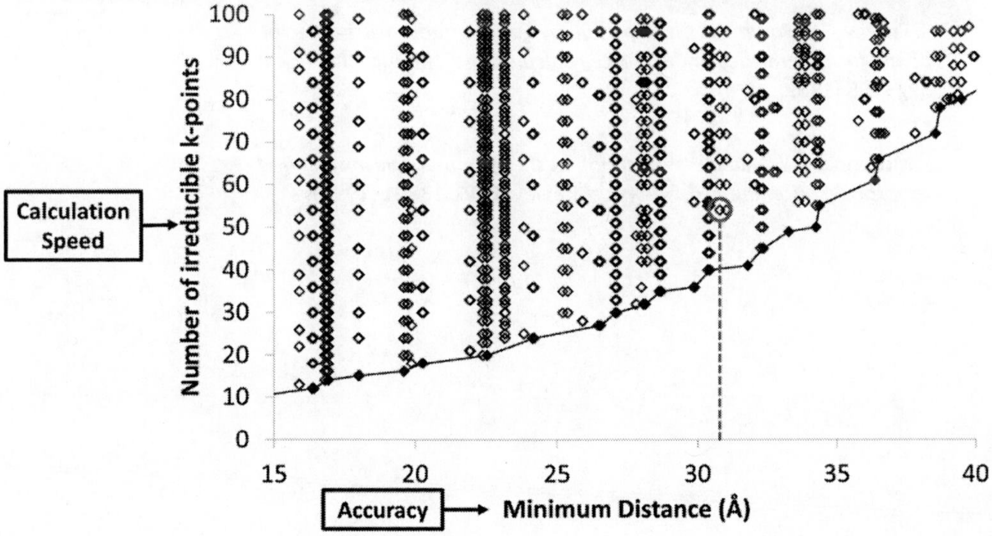

Pareto Frontier

JOHNS HOPKINS
WHITING SCHOOL of ENGINEERING

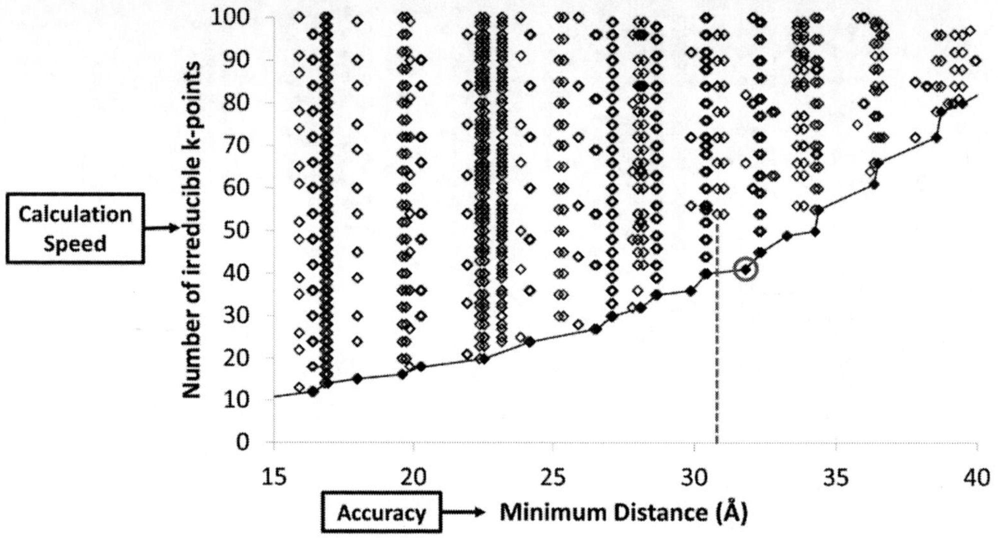

About 25 years ago…

JOHNS HOPKINS
WHITING SCHOOL of ENGINEERING

S. Froyen, "*Brillouin-zone integration by Fourier quadrature: Special points for superlattice and supercell calculations*", Physical Review B **39**, 3168 (1989).

J. Moreno and J. M. Soler, "*Optimal meshes for integrals in real- and reciprocal-space unit cells*" Physical Review B **45**, 13891 (1992).

About 25 years ago…

Mesh type	N	$\mathbf{b}_0$	N_{BZ}	N_{IBZ}	$(2l_{cut}/A_0)^2$	N_{star}	
sc	2	p_0, p_2	1	1	1	1	
bcc	2	p_0	2	2	2	2	
fcc	2	p_0, p_1, p_2	4	2	3	3	
sc	4	p_2	8	1	4	4	*
bcc	4	p_0	16	5	8	7	
sc	6	p_0, p_2	27	4	9	8	
fcc	4	p_1, p_2	32	4	12	11	
sc	8	p_2	64	4	16	14	*
bcc	6	p_0	54	8	18	16	*
sc	10	p_0, p_2	125	10	25	22	
fcc	6	p_0, p_1, p_2	108	10	27	24	
bcc	8	p_0	128	14	32	27	
sc	12	p_2	216	10	36	31	*
fcc	8	p_1, p_2	256	16	48	41	*
sc	14	p_0, p_2	343	20	49	42	
bcc	10	p_0	250	20	50	43	
sc	16	p_2	512	20	64	54	*
bcc	12	p_0	432	30	72	61	
fcc	10	p_0, p_1, p_2	500	28	75	64	*
sc	18	p_0, p_2	729	35	81	69	
bcc	14	p_0	686	40	98	83	
sc	20	p_2	1000	35	100	85	*
fcc	12	p_1, p_2	864	40	108	92	*
sc	22	p_0, p_2	1331	56	121	102	
bcc	16	p_0	1024	55	128	107	*
sc	24	p_2	1728	56	144	121	*
fcc	14	p_0, p_1, p_2	1372	60	147	124	*

J. Moreno and J. M. Soler, "*Optimal meshes for integrals in real- and reciprocal-space unit cells*" Physical Review B **45**, 13891 (1992).

Method Usage Frequency

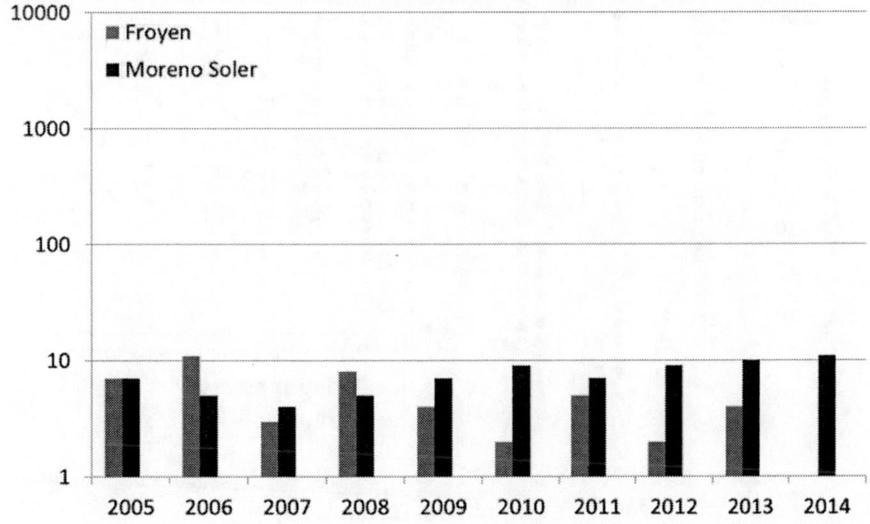

Monkhorst-Pack is more commonly used

JOHNS HOPKINS
WHITING SCHOOL of ENGINEERING

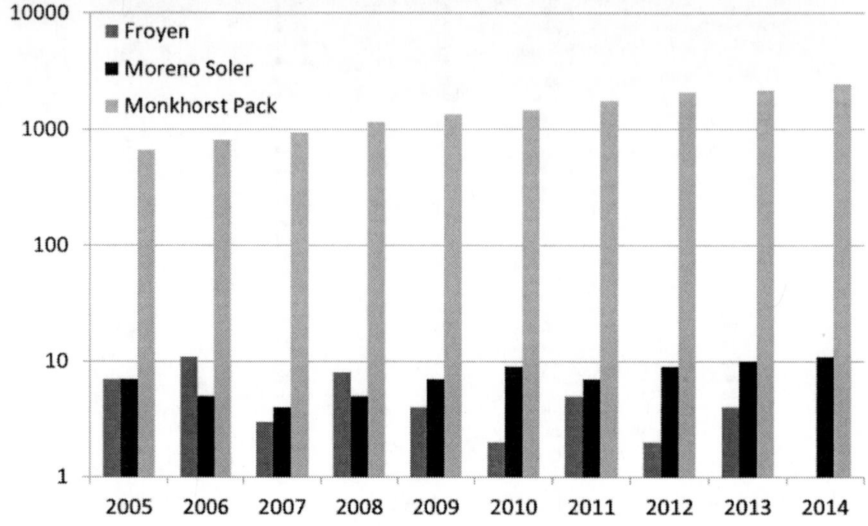

Pareto Frontier

JOHNS HOPKINS
WHITING SCHOOL of ENGINEERING

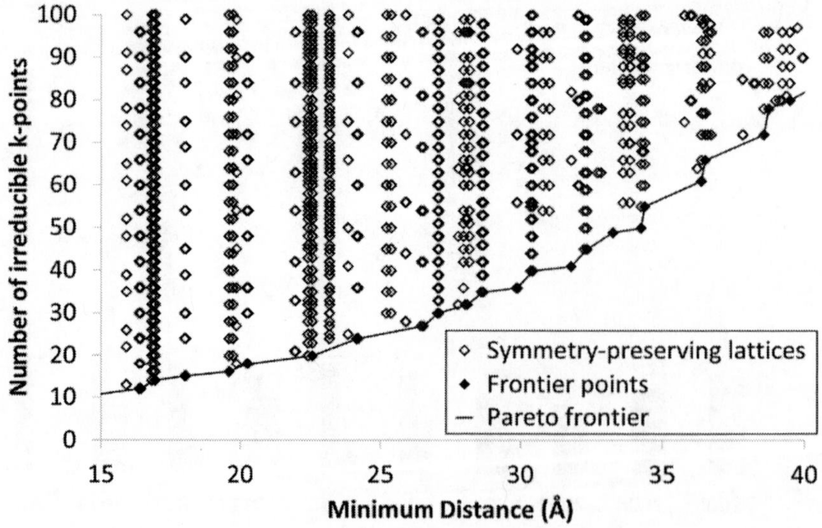

Pareto Frontier

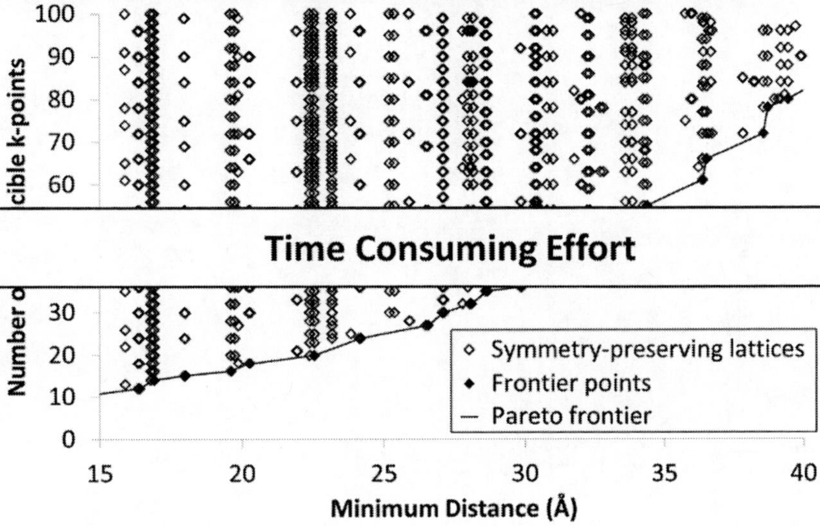

K-Point Grid Database

Johns Hopkins
Whiting School of Engineering

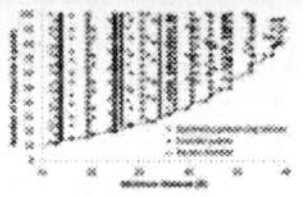

K-Point Grid Database

JOHNS HOPKINS
WHITING SCHOOL of ENGINEERING

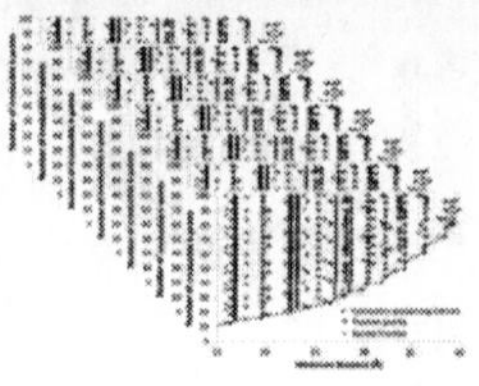

K-Point Grid Database

JOHNS HOPKINS
WHITING SCHOOL of ENGINEERING

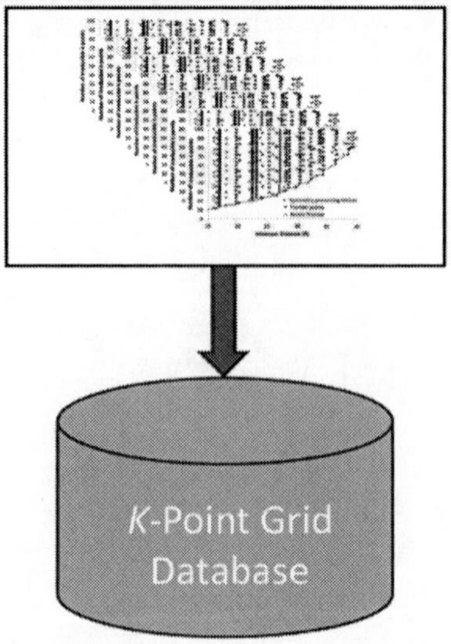

K-Point Grid Database

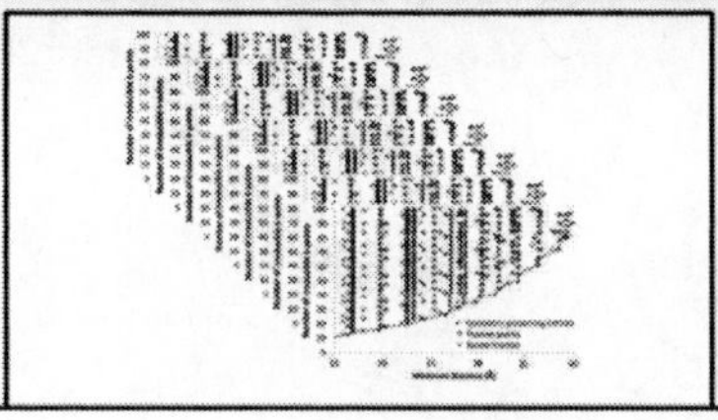

389,579 total grids, supporting up to 157,464 total *k*-points per grid

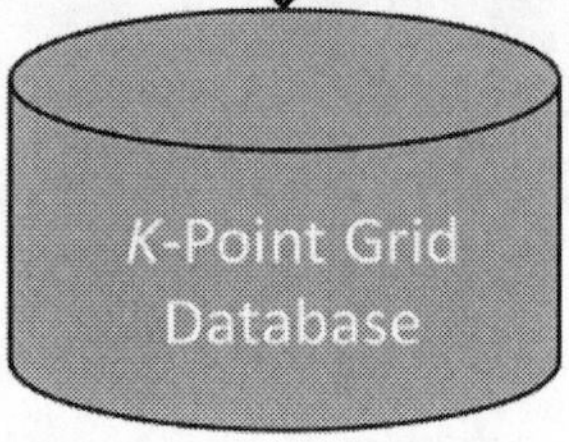

Benchmark on 102 materials for convergence within 1 meV / atom

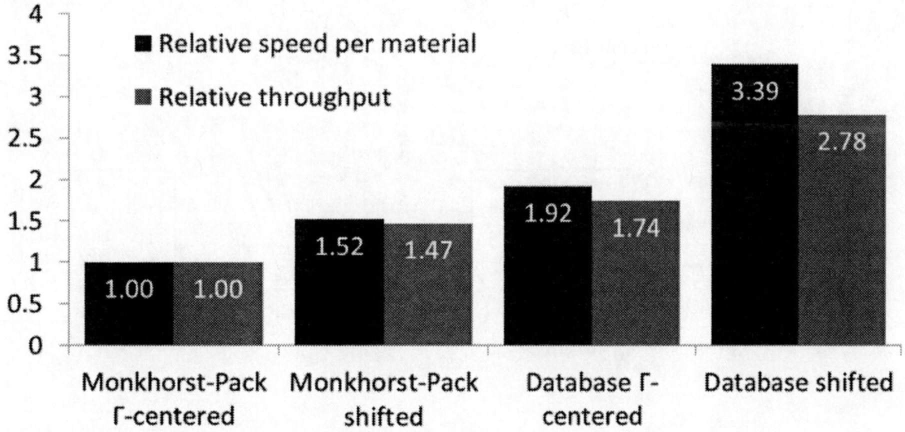

Choosing the minimum distance

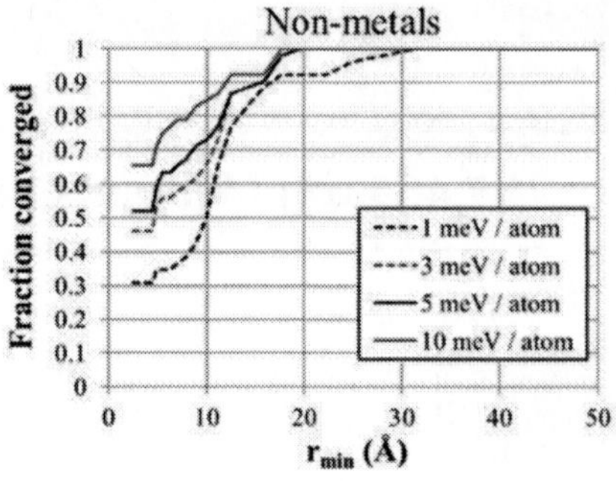

Choosing the minimum distance

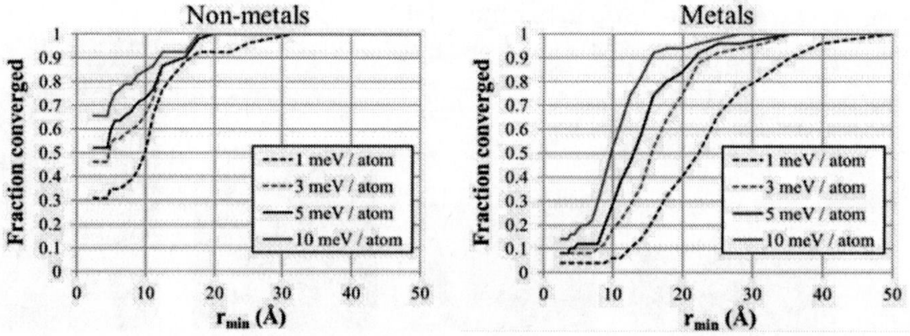

Choosing the minimum distance

JOHNS HOPKINS WHITING SCHOOL of ENGINEERING

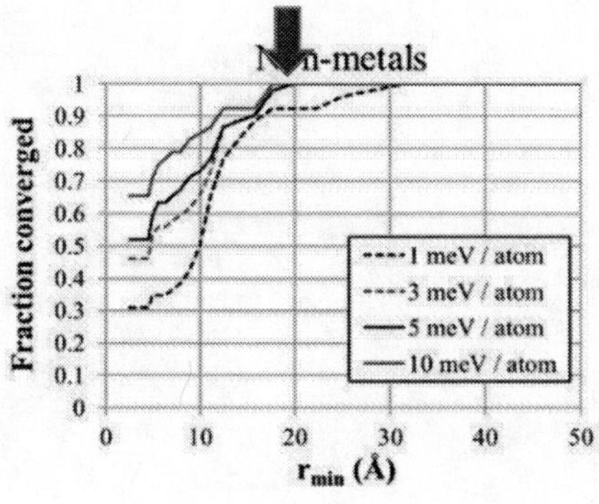

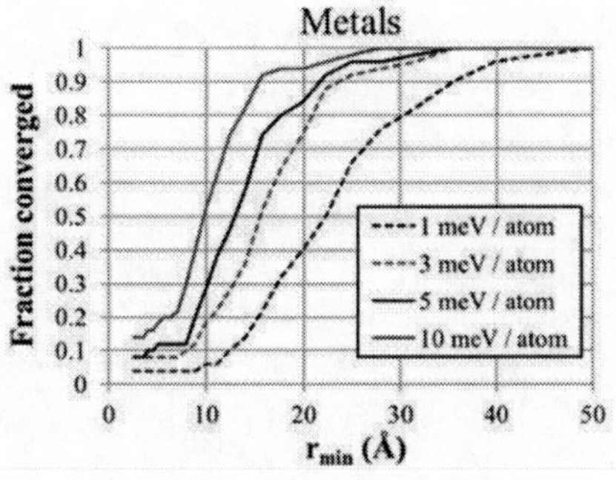

Choosing the minimum distance

JOHNS HOPKINS WHITING SCHOOL of ENGINEERING

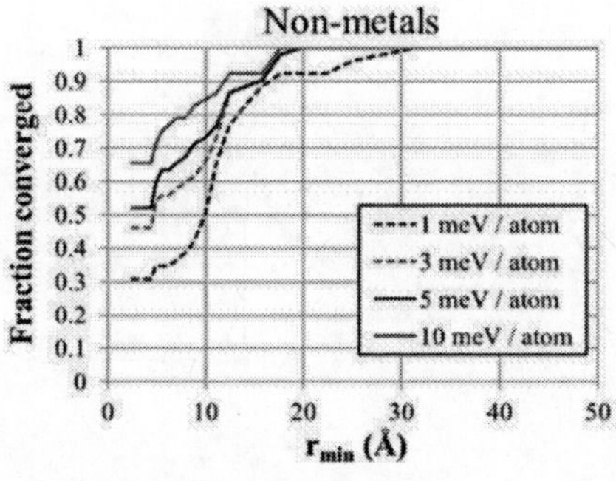

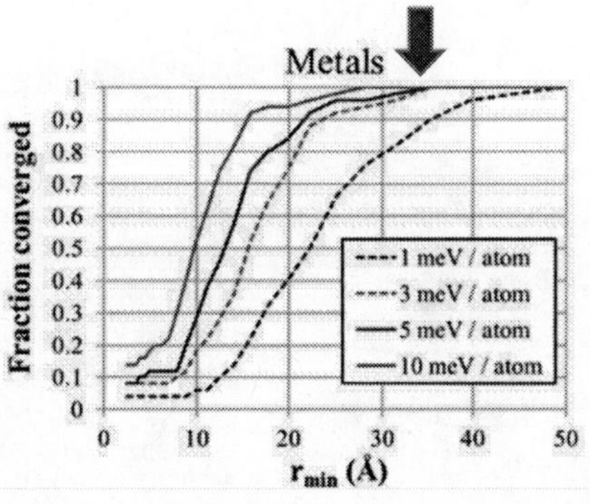

Choosing the minimum distance

JOHNS HOPKINS WHITING SCHOOL of ENGINEERING

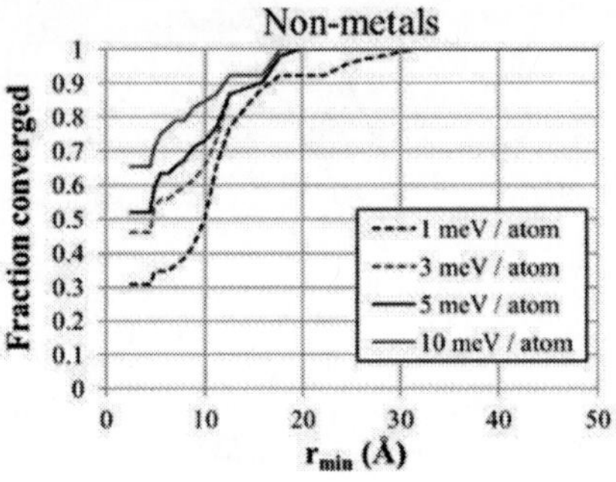

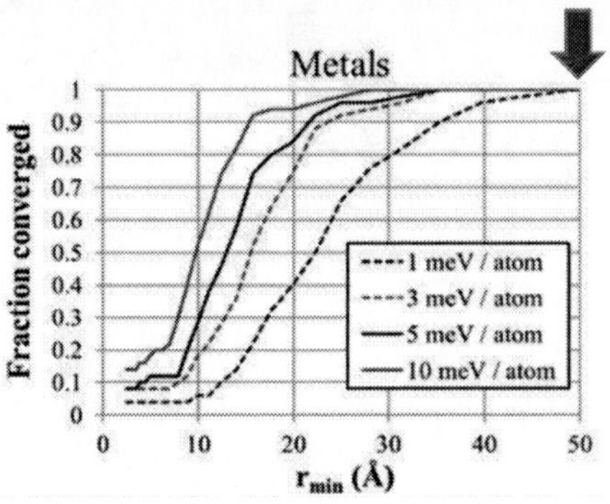

Choosing the minimum distance

JOHNS HOPKINS WHITING SCHOOL of ENGINEERING

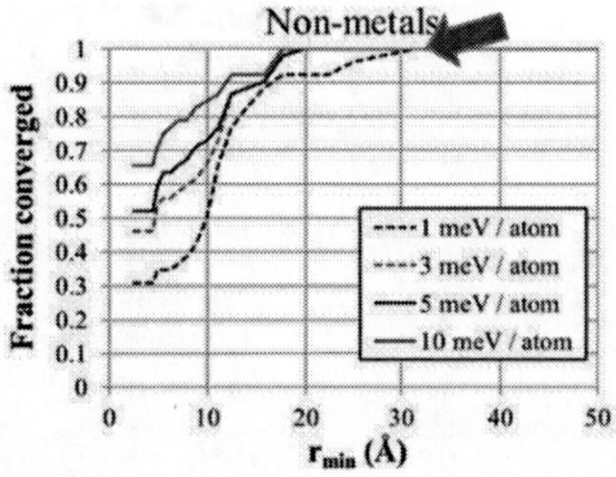

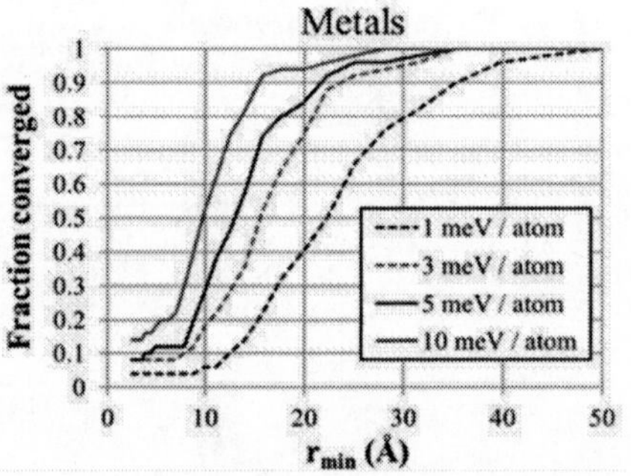

Choosing the minimum distance

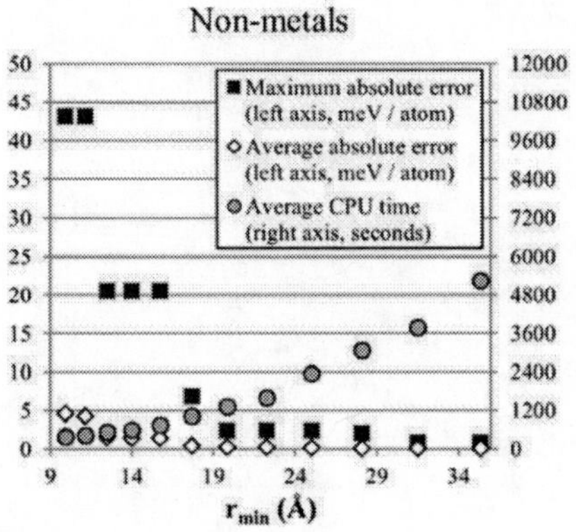

Choosing the minimum distance

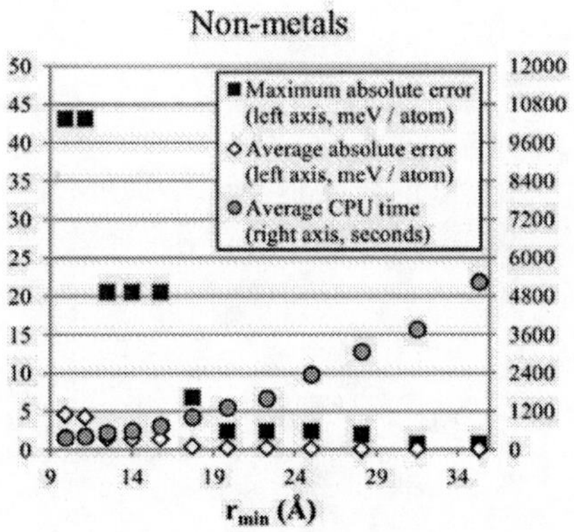

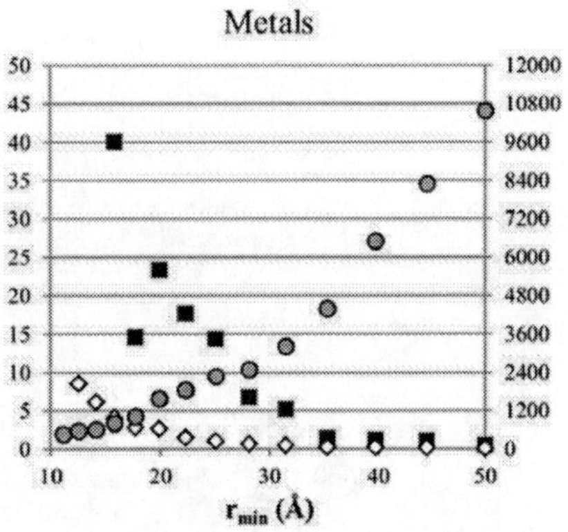

Choosing the minimum distance

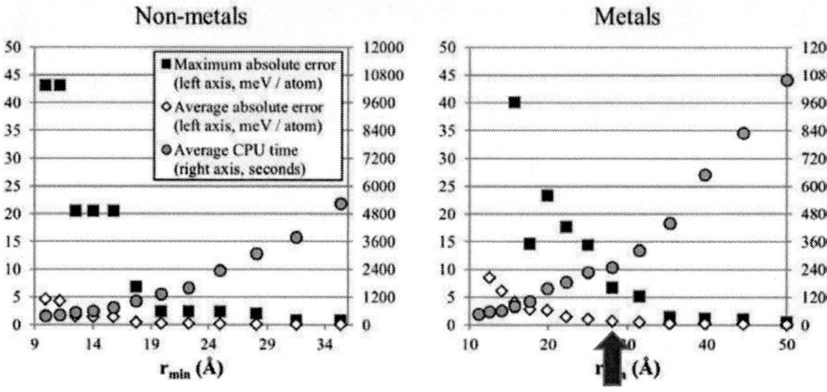

Choosing the minimum distance

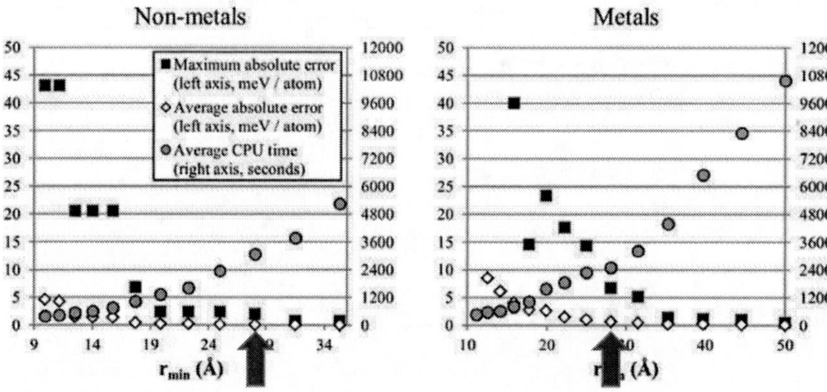

Choosing the minimum distance

JOHNS HOPKINS
WHITING SCHOOL of ENGINEERING

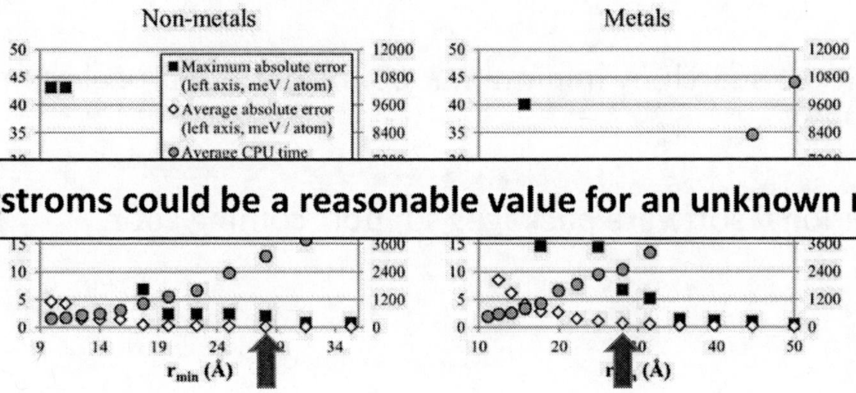

28.1 Angstroms could be a reasonable value for an unknown material.

Convenience

JOHNS HOPKINS
WHITING SCHOOL of ENGINEERING

- No installation required (download a simple Bash script).
- Single convergence parameter.
- The same input file can be used for primitive cell, supercells, surfaces, etc.
- Choice of primitive cell vectors does not change *k*-point lattice.
- Currently works with VASP software package.

Additional features

JOHNS HOPKINS
WHITING SCHOOL of ENGINEERING

- Auto-detection of slabs, nanowires, and nanoparticles.
- Ability to search by minimum *k*-points per reciprocal atom (KPPRA).
- Additional software packages support coming soon.

More Information

JOHNS HOPKINS
WHITING SCHOOL of ENGINEERING

http://muellergroup.jhu.edu/K-Points.html

470116 Machine Generated Coarse-Grained Force-Fields for Efficiently and Effectively Addressing Transferability

Abstract:

A key objective of the Materials Genome Initiative (MGI) is to create a coherent framework to perform coarse-grained (CG) simulations in a streamlined and efficient manner to reduce the time and cost of novel material design and discovery. In order to gain more widespread application in that endeavor, it is imperative to make progress on one of the main limitations of coarse-grained potentials and efficiently and effectively address the issue of transferability. In this study, we develop a coarse-grained potential for polystyrene at a resolution of one CG site per monomer using Iterative Boltzmann Inversion (IBI) in conjunction with Uncertainty Quantification (UQ) techniques. We also introduce a new Spectral Monte Carlo (SMC) method to calculate the necessary distribution functions for the IBI framework which is orders of magnitude more efficient than typical histogram methods. Using a minimal set of IBI-parameterized CG force-fields derived across a range of temperatures and pressures (which constitutes an initial guess), Bayesian methods are subsequently applied to efficiently derive machine generated CG force-fields which are accurate at arbitrary state points. Multiple properties including density, cohesive energy density, and compressibility are used to calibrate the Bayesian derived potential. We show the algorithm is able to accurately reproduce the calibration properties across the entire range of state points, and examine the effect of the machine generated force-field on other more problematic coarse-grained properties such as diffusion and modulus. We conclude with an outlook at how these analyses can applied to other properties in more complicated material systems to better understand the boundaries of applicability of these CG potentials.

472573 Development of an Inter-Atomic Potential for Molecular Dynamics Simulations of Stanene Using a Genetic Algorithm Based Framework

Abstract:

The growth of single layer tin (stanene) on a Bi_2Te_3 substrate has engendered a great deal of interest, in part due to staneneâ€™s predicted exotic properties. In particular, stanene has attracted lot of attention owing to its tremendous promise in topological insulation, large-gap 2D quantum spin hall states, lossless electrical conduction, enhanced thermoelectricity, and topological superconductivity. Most of the previous work on stanene has focused on its electronic properties. Atomistic investigations of growth mechanisms (needed to guide synthesis), phonon transport (crucial for designing thermoelectrics), and thermo-mechanical behavior of stanene are scarce. This paucity is primarily due to the lack of inter-atomic potentials that can accurately capture atomic interactions in stanene. To address this, we have developed a bond-order potential (BOP) based on Tersoffâ€™s formalism that can accurately capture bond breaking/formation events, structure, energetics, thermodynamics, phonon frequencies, thermal conductivity, and mechanical properties of single layer tin. We determine the BOP parameters by fitting to a training dataset containing (a) structure, (b) equation of state (energy vs area), (c) elastic constants, and (d) phonon dispersion of stanene obtained from our density functional theory calculations. To optimize this potential, we employed a hybrid global optimization scheme based on genetic algorithms and Nelder-Mead simplex. Finally, we employed our newly developed BOP to study anisotropy in thermal conductivity of stanene sheets, temperature induced rippling, as well as dependence of anharmonicity and thermal conductivity on temperature.

473668 A Lattice Kinetic Monte Carlo Method for Advective Systems

Abstract:

The lattice kinetic Monte Carlo (LKMC) method is a particularly appealing mesoscopic technique for simulating spatially-resolved particulate systems because of its inherent simplicity, computational efficiency, and versatility. However, the vast majority of LKMC simulations have been applied towards problems in which diffusion is the sole transport mechanism, i.e., in which external flow is absent. Example applications are extremely diverse, ranging from defect diffusion in metals and semiconductors [1], to crystal growth [2], to thrombosis in blood flow [3]. Recently, Flamm *et al.* [4,5] introduced an LKMC algorithm for including advective particle transport in a prescribed fluid flow field. In this technique, the advective force exerted on each particle due to fluid flow was used to augment particle hopping rates in the direction of the fluid flow. The algorithm was shown to predict accurately the temporal evolution of a collection of particles under strongly non-equilibrium conditions, including two-dimensional Taylor dispersion in a straight channel and in an expansion flow. However, the method was not evaluated under equilibrium, or weakly non-equilibrium, conditions where adherence to detailed balance becomes relevant.

Here, we develop and analyze new rate expressions for advective LKMC for particulate flow situations. Using equilibrium and non-equilibrium test problems, we show that the new algorithm is able to correctly predict particle dynamics under general flow conditions, while also satisfying detailed balance. In the equilibrium case, the method is analyzed in the context of biased particle diffusion in a non-linear potential field, and the results compared to those from Metropolis Monte Carlo simulations. Using this comparison, we show that the new technique exactly satisfies detailed balance. Next, Taylor dispersion simulations are used to probe the performance of the method in a non-equilibrium setting. Using Brownian dynamics (BD) simulations as a reference, we demonstrate a relationship between the LKMC lattice spacing and the BD timestep. This relationship is used to derive formal error estimates for LKMC simulation of advection-diffusion problems. We conclude by discussing the relative computational advantages of BD and LKMC.

[1] J. Dai, J. M. Kanter, S. S. Kapur, W. D. Seider, and T. Sinno, Phys. Rev. B **72**, 134102 (2005).

[2] A. C. Levi and M. Kotrla, J. Phys.: Condens. Matter **9**, 299 (1997).

[3] M. H. Flamm et al., Blood **120**, 190 (2012).

[4] M. H. Flamm, S. L. Diamond, and T. Sinno, J. Chem. Phys. **130**, 094904 (2009).

[5] M. H. Flamm, T. Sinno, and S. L. Diamond, J. Chem. Phys. **134**, 034905 (2011).

454069 Continuous Fractional Component Monte Carlo in the Gibbs Ensemble

Abstract:

Continuous Fractional Component Monte Carlo in the Gibbs Ensemble

Ali Poursaeidesfahani, Ariana Torres-Knoop, David Dubbeldam, Thijs J.H. Vlugt

We introduced a new formulation of the fractional component Monte Carlo approach in the Gibbs ensemble [1]. In our formulation, only a single fractional particle per component is used, which has three main advantages: (1) chemical potentials of the two boxes are calculated automatically (without having to resort to a separate Widom's test particle computation), and we have shown analytically that the derived expression for the chemical potential is mathematically identical to that in the Gibbs ensemble; (2) independent biasing for each simulation box improves sampling of the configurational space; (3) the maximum allowed changes in the coupling parameter can be chosen differently for each box. Therefore, larger changes in the value of the coupling parameter can be used when the fractional particle is located in the box corresponding to the gas phase compared to the coupling used in the liquid phase. In addition to the trial-moves required to thermalize the system, the proposed method involves three additional trial-moves to satisfy the requirements of equal chemical potential and pressure of the gas and liquid phase in the Gibbs ensemble: (1) attempts to change the coupling parameter of the fractional particle within the range [0,1]; (2) trial-moves to swap the fractional particle from one simulation box to the other one; (3) attempts to change the fractional particle into an integer particle, and at the same time, change a randomly selected integer particle in the other simulation box into a fractional particle. Our method has been tested for the system of Lennard-Jones particles and the TIP3P-Ew water model at different system sizes and temperatures. Chemical potentials and densities calculated for two boxes are compared with values

computed using the conventional Gibbs ensemble with Widom's test particle method. Excellent agreement was found for both densities and chemical potentials. The acceptance probability for the particle exchange has improved significantly in our formulation compared to the conventional Gibbs ensemble (typically from 2% to 40% for LJ particles, and from 0.6% to 8.5% for simulations with TIP3P-Ew water model). We found that the method is superior to conventional MC, but also significantly better than state-of-the-art Configurational-bias MC methodology. It also represent a significant improvement over a previous formulation of CFCMC in the Gibbs ensemble, both in acceptance probabilities and performance as well as from a theoretical point of view. We have also addressed the question of including or ignoring the contribution of the fractional particle in calculation of ensemble averages. We have shown analytically that one should disregard the fractional particle while computing the average density of the system.

[1] Poursaeidesfahani A., Torres-Knoop A., Dubbeldam D, Vlugt T. J.H., **J. Chem. Theory Comput., 2016, 12,** 1481-1490

462826 Improved Coarse-Grained Models through Local Density Potentials Optimized with the Relative Entropy

Abstract:

Bottom-up multiscale techniques are often used to develop coarse grained (CG) models of atomistic molecular simulations to probe large length and time scales. Conventionally, CG models use pair potentials for nonbonded forces to capture the renormalization of detailed interactions of a reference all-atom system. Such pair potentials, while computationally attractive, neglect the inherently multibody contributions of the local environment of a CG site to its energy, which is inevitably embedded in the CG model as a result of integrating over removed degrees of freedom. This makes CG models highly sensitive to the thermodynamic state point (temperature, density, concentration, etc.) of the all-atom reference from which they are parameterized, and as such, limits their transferability to other state points. Here, we propose to improve CG models through computationally fast many body potentials, in which multibody energies depend in a mean-field manner on local densities of different atomic species and serve as additive corrections to conventional CG pair potentials. These local density potentials have the same mathematical structure as embedded-atom or bond-order models designed to capture multibody electronic effects in metallic systems. We show that the relative entropy coarse-graining framework offers a systematic route to parameterizing local density potentials. We explore this approach in the development of implicit solvation strategies for aqueous solvation of hydrophobic solutes. CG models of solutes and macromolecules that represent solvent implicitly through effective intra-solute interactions offer tremendous potential speed-ups through the elimination of explicit solvent molecules. Our results reveal that the local density approach enables the models to capture higher order cooperativity inherent to water-mediated hydrophobic interactions, which results in better transferability of local density augmented implicit solvent CG forcefields. More generally, this approach can be easily extended to arbitrary CG models of systems beyond solvation, e.g. phase behavior, where density and concentration transferability play an important role.

459284 Explicit Nonlinear Collective Variables and Biased Molecular Dynamics Using Autoencoders

Abstract:

Macromolecular and biomolecular folding landscapes typically contain high free energy barriers that impede efficient sampling of the thermally accessible configurational space by conventional molecular dynamics simulation. Biased sampling approaches (e.g., umbrella sampling, metadynamics, adaptive biasing force) seek to improve sampling by artificially driving the simulation along some collective variables (CVs) to accelerate exploration of configurational space. The success of these methods critically depends on the availability of good CVs that are associated with the important collective dynamical transitions, span the accessible phase space, and do not contain latent free energy barriers. Nonlinear dimensionality reduction techniques (e.g., Laplacian eigenmaps, diffusion maps, Isomap, locally linear embedding) can systematically identify appropriate CVs as nonlinear combinations of molecular degrees of freedom corresponding to the collective molecular motions. A critical impediment to implementing these CVs within biased molecular dynamics simulations is the unavailability of an explicit functional relationship between the CVs and atomic coordinates, which precludes the mapping of biasing forces in the CVs to real-space forces on the atoms. Innovative resolutions of this fundamental difficulty have been proposed by using the CVs to initialize unbiased simulations in undersampled regions[1], construct approximate parameterizations of the CVs in basis functions centered on representative molecular configurations[2], employ restriction/lifting and coarse-grained projection between the molecular and CV descriptions[3], and perform biased sampling in proxy physical variables correlated with the collective CVs[4].

In this work, we report the use of autoassociative artificial neural networks (autoencoders) to learn low-dimensional nonlinear subspaces containing the collective dynamical motions of biomolecular simulations and furnish nonlinear CVs that are explicit functions of the atomic coordinates. We have integrated this data-driven CV discovery with

umbrella sampling within the molecular simulation package OpenMM[5] to perform accelerated sampling directly in the low-dimensional subspace by propagating the CV biasing forces into real-space forces on the atoms. By interleaving successive rounds of CV discovery and biased sampling, we have established an approach to iteratively discover and refine the CVs and efficiently sample the thermally accessible phase space by parsimonious biasing along the important collective molecular motions. We describe applications of our approach to systematically discover good collective variables and efficiently calculate free energy surfaces for alanine dipeptide and the Trp-cage miniprotein.

1. J. Preto and C. Clementi *Phys.Chem.Chem.Phys.* **16** 19181 (2014)

2. B. Hashemian, D. Millán, and M. Arroyoa *J. Chem. Phys.* **139** 214101 (2013)

3. T.A. Frewen, G. Hummer, and I.G. Kevrekidis *J. Chem. Phys.* **131** 134104 (2009)

4. A.L. Ferguson, A.Z. Panagiotopoulos, P.G. Debenedetti and I.G. Kevrekidis *J. Chem. Phys.* 134 135103 (2011)

5. P. Eastman *et al. J. Chem. Theor. Comput.* **9** 1 461-469 (2013)

467554 Solvent Activity in Electrolyte Solutions By Molecular Simulation of the Osmotic Pressure

Abstract:

Describing electrolyte solutions by molecular modelling and simulation is a promising approach for characterizing structural and thermodynamic effects of electrostatic interactions in a self-consistent way [1-3]. The solvent activity is one of the most important properties of an electrolyte solution. It is directly related to the the vapour pressure and the freezing point, and it is connected to the activity coefficient of the salt which controls the solubility limit.

However, a reliable determination of the solvent activity in electrolyte solutions by molecular simulation is challenging numerically. In the present work, a novel method for determining the solvent activity is presented. Thereby, the electrolyte solution is simulated in contact with the pure solvent. Between the two phases, there is a virtual membrane which is permeable only for the solvent. This is implemented by an external field which acts on the solutes only and confines the solute to a part of the simulation volume. The osmotic pressure can then be computed with a high accuracy from the force acting on the membrane, so that reliable data on the solvent activity are obtained [4].

The present method is validated and applied to aqueous electrolyte solutions for which results for the solvent activity are available from other recent studies [1, 2]. The simulations are performed with an extended version of the molecular simulation program ms2, which is freely available for academic users [5]. A good agreement with the literature is observed. On this basis, a systematic investigation of aqueous alkali halide salt solutions is carried out. Even though the employed series of ion models was developed without considering activity data for the model parameterization, which was mainly based on density data [3], a good agreement with experimental solvent activity data is found in many cases.

[1] F. Moucka, I. Nezbeda, and W. R. Smith, Journal of Chemical Physics 139 (2013) 124505.

[2] Z. Mester and A. Z. Panagiotopoulos, Journal of Chemical Physics 142 (2015) 044507.
[3] S. Reiser, S. Deublein, J. Vrabec, and H. Hasse, Journal of Chemical Physics 140 (2014) 044504.
[4] M. Kohns, S. Reiser, M. Horsch, and H. Hasse, Journal of Chemical Physics 144 (2016) 084112.
[5] C. W. Glass, S. Reiser, G. Rutkai, S. Deublein, A. Köster, G. Guevara Carrión, A.Wafai, M. Horsch, M. Bernreuther, T. Windmann, H. Hasse, and J. Vrabec, Computer Physics Communications 185 (2014) 3302-3306.

465348 Simulation Methods to Probe RNA Conformational Dynamics

Abstract:

Since the seminal discovery that Ribonucleic acid (RNA) molecules can catalyze complex reactions similar to proteins, their functional repertoire has dramatically expanded with fundamental roles in many regulatory processes such as gene expression and control. RNA therefore is no longer considered a passive information carrier, but rather a molecule with versatile chemical and physical properties that can be exploited in therapeutic and engineering applications. Underlying such applications, one ubiquitous and critical physical attribute is dynamics encompassing a large number of modes and spanning vast spatiotemporal scales. Dynamics in RNA are intimately connected to intrinsic structural features (e.g. base-pairing) as well as extrinsic factors (e.g. ligand binding). While equilibrium fluctuations in biomolecules such as RNA are inherent, it is the large-scale conformational transitions that play a dominant role in interfacial processes such as biomolecular recognition. The information on such conformational states is encoded in the physical free energy landscapes of RNA molecules, the robust sampling of which can provide key thermodynamic and kinetic information on functional conformational ensembles. In this talk, we will present applications of new and emerging enhanced sampling molecular simulation methods based upon all-atom and structure-based models of small RNA molecules that are known to undergo large-scale conformational changes. A specific example that will be highlighted is of the transactivation response element RNA (TAR-RNA) from the genome of HIV-1 virus possessing versatile conformational states that are rapidly becoming targets of novel therapeutics.

472174 Protracted Colored Noise Dynamics in Molecular Dynamics Simulations of Block Copolymers

Abstract:

Block copolymers (BCPs) can phase separate to form periodic structures with small spacings, making BCPs an attractive option for furthering the ability of optical lithography to make dense arrays of features. Simulation of structural formation in these systems can help optimize the BCP system to produce small repeatable structures. Unfortunately, these BCP photoresist systems are highly viscous and the simulation of order formation requires very long simulation times. While coarse-grain and mean-field approaches have been applied to increase efficiency of these simulations, we apply a stochastic simulation to a coarse-grain model to further increase efficiency. This method, Protracted Colored-Noise Dynamics (PCND), adds colored-noise forces to increase local diffusivity while maintaining constant temperature with an integral energy controller. This method has been adapted to polymers by selectively applying the stochastic force to the polymer contour to artificially increase reptation to increase the efficiency of viscous polymer simulations. This work demonstrates the effects of PCND on a coarse-grained BCP system and explores the primary parameters involved in PCND. We show that PCND is able to increase the kinetics of defect annihilation in BCP systems more effectively than raising the temperature. It also avoids the unrealistic perturbation of polymer density and conformation that accompanies large temperature increases. PCND is shown to alter some properties of the BCP such as the pitch, yet once PCND is turned off the system returns to its natural state.

462958 Efficient Simulation of the Hydrophobic Effect in Biological Self-Assembly

Abstract:

The assembly of proteins into complexes and higher-order structures occurs on the scale of nanometers, where the thermodynamics of solvation are not well described by macroscopic thermodynamics and the microscopic structure of water determines the hydration layer around a protein. In atomistic as well as coarse-grained simulations, solvent is often taken into account explicitly to describe the hydrophobic effect and protein surface interactions, which severely limits the time scales accessible in simulation because of integrating the water degrees of freedom. Inspired by recent work [1], we propose a novel implementation of an implicit solvation scheme in Molecular Dynamics on GPUs, which is based on the theory for the hydrophobic effect of Lum, Chandler and Weeks for the nanometer-scale solvation and a smoothed particle description on larger scales [2]. This model captures the hydrophobic effect due to exclusion of solvent fluctuations from small and large cavities accurately, while being computationally efficient. The goal is to routinely do simulations on millisecond time scales in biological self-assembly on small clusters of GPUs.

[1] S. Vaikuntanathan, G. Rotskoff, A. Hudson, and P. L. Geissler, "Necessity of capillary modes in a minimal model of nanoscale hydrophobic solvation," *Proc. Natl. Acad. Sci.*, vol. 113, no. 16, pp. E2224, 2016.

[2] H. Lei, C. J. Mundy, G. K. Schenter, and N. K. Voulgarakis, "Modeling nanoscale hydrodynamics by smoothed dissipative particle dynamics," *J. Chem. Phys.*, vol. 142, no. 19, p. 194504, 2015.

467924 First Principles Monte Carlo Simulations of Reaction Equilibria in Compressed Vapors

Abstract:

Predictive modeling of reaction equilibria presents one of the grand challenges in the field of molecular simulation. Difficulties in the study of such systems arise from the need (i) to accurately model both strong, short-ranged interactions leading to the formation of chemical bonds and weak interactions arising from the environment, and (ii) to sample the range of time scales involving frequent molecular collisions, slow diffusion, and infrequent reactive events. Here we present a novel reactive first principles Monte Carlo (RxFPMC) approach that allows for investigation of reaction equilibria without the need to pre-specify a set of chemical reactions and their ideal-gas equilibrium constants. We apply RxFPMC to investigate a nitrogen/oxygen mixture at $T = 3000$ K and $p = 30$ GPa, i.e., conditions that are present in atmospheric lightning strikes and explosions. The RxFPMC simulations show that the solvation environment leads to a significantly enhanced NO concentration that reaches a maximum when oxygen is present in slight excess. In addition, the RxFPMC simulations indicate the formation of NO2 and N2O in mole fractions approaching 1%, whereas N3 and O3 are not observed. The equilibrium distributions obtained from the RxFPMC simulations agree well with those from a thermochemical computer code parameterized to experimental data.

468588 Replica-Exchange on-the-Fly Parameterization: Application of a High-Precision Free-Energy Method to Understanding the Roles of the M129V/D178N Polymorphisms in the Conformational Thermodynamics of Human Prion Protein

Abstract:

Barriers in the sampling of "hidden" variables constitute the most important limitations on accuracy and precision of free-energy landscapes computed using collective-variable (CV) biasing approaches, such as thermodynamic integration, umbrella sampling, Metadynamics, adaptive-biasing force calculations, and others. So-called orthogonal-space sampling variants of these methods attack this issue using additional biasing, but they aren't always guaranteed to find all relevant hidden variables. This problem is particularly acute in situations where one wishes to understand changes in protein conformational thermodynamics induced by substitutions of amino acid side-chains that evidently do not result in changes to any specific interactions (no net loss or gain of H-bonds, salt-bridges, dipole-dipole interactions, etc.). A prominent example is the M/V polymorphism at position 129 of the human prion protein (hPrP), which has recently been shown to be a major determinant of a variety of histologically and clinically distinct phenotypes of transmissible spongiform encephalopathies (TSEs). We present here a new free-energy method termed Replica-Exchange On-the-Fly Parameterization (REOTFP), which combines the advantages of history-free CV-biasing of temperature-accelerated MD with the general hyperthermal enhanced sampling of replica-exchange MD, to overcome the hidden-variables problem. We show that REOTFP can produce free-energy profiles precise enough to rationalize effects of the M/V-129 polymorphism of hPrP. In particular, we use REOTFP to measure the thermodynamic stability of the N-terminal β-sheet of native hPrP and show that V129 causes loss of a low-stability metastable state that may correlate to a phenotype protective against TSE acquisition. Additionally, in the presence of the heritable-TSE-associated mutation D178N, REOTFP predicts that V129 causes acquisition of this metastable intermediate, which may underlie the distinct natures of inherited Creutzfeldt-Jakob disease and Fatal Familial Insomnia, two different TSE's that seem to owe their divergence solely to the M/V polymorphism at position 129. We compare the performance of REOTFP to both serial OTFP and well-tempered Metadynamics and explain the basis of REOTFP's superiority in terms of specific hidden variables that it efficiently samples with no additional bias. On a per-unit-error basis, REOTFP is vastly less expensive than virtually any serial free-energy method to date, and should therefore open new pathways of inquiry previously off-limits to existing free-energy methods.

456118 Molecular Dynamics Simulation of the Thermodynamics of Bismuth Telluride Exfoliation in Ionic Liquids

Abstract:

Bismuth Telluride (Bi_2Te_3) is a topological insulator that possesses unique thermoelectric properties at near room temperatures, and it is a valuable material for waste heat recovery and other niche applications.[1] The crystal structure of Bi_2Te_3 contains quintuple sheets of Te1-Bi-Te2-Bi-Te3 bonded covalently with a thickness of approximately 1 nm. The bulk Bi_2Te_3 material is composed of these quintuple sheets stacked together with only dispersion interactions acting between the neighboring nanosheets. The thermoelectric performance of Bi_2Te_3 can be improved by separating the bulk material into two-dimensional (2D) nanosheets. When these materials are exfoliated into very thin layers, increased phonon scattering decreases the thermal conductivity, while their high crystallinity enhances their electrical conductivity. As a result, these materials possess high crystallinity, unaffected topological surface states, and high anisotropy, providing valuable thermoelectric properties.[2, 3]

In order to synthesize Bi_2Te_3 nanosheets, several different methods have been reported, such as vapor-phase decomposition[4] and liquid-phase synthesis.[1] Liquid-phase synthesis is preferred because of its low cost and simplicity. Ionic liquids (ILs) have been proposed as environmentally-friendly solvents and have previously been used to exfoliate layered materials such as graphene[5] and Bi_2Te_3.[6] ILs provide some unique advantages (versus volatile organic solvents), due to their low vapor pressure, high thermal stability, high ionic conductivity, and low melting points.

Molecular Dynamics (MD) simulations have been widely used to provide atomistic insight into the thermophysical properties of ILs and their interfacial interactions.[7] In addition, MD simulations have been successfully used to model the exfoliation of bilayer graphene,[5] hexagonal boron nitride (h-BN),[8] and recently Bi_2Te_3.[6] However, there is still a considerable lack of knowledge about how to design the most effective IL for exfoliating 2D nanomaterials. There must be a balance of interactions that allow an IL solvent to initialize the exfoliation mechanism and stabilize the individual nanosheets; this mechanism involves both dispersion and electrostatic interactions. For instance, while Bi_2Te_3 nanosheets are charge-neutral, Te atoms cover the basal plane of the quintuple sheets, which presents a negatively charged surface. Currently, these solid-liquid interfacial interactions between IL solvents and chalcogenide materials are not well characterized, and a clearer molecular-level description is needed.

In our study, we use MD simulations to model the initial exfoliation mechanisms of Bi_2Te_3 in several different imidazolium-based ILs. These imidazolium-based ILs have surface energies estimated to be 63-83 mJ/m^2,[9] which provide a commensurate match with the estimated surface energy of Bi_2Te_3 (63-81 mJ/m^2)[6]. Eight different ILs are modeled, containing [C_4mim^+] and [C_2mim^+] as cations and [Tf_2N^-], [PF_6^-], [Br^-], and [Cl^-] as anions, using steered molecular dynamics (SMD) and the weighted histogram analysis method (WHAM). With these different solvents, several different exfoliation mechanisms are explored (peeling, pulling, and sliding), and the thermodynamics of the different exfoliation routes are quantified. Our simulations indicate that the peeling mechanism is the most probable mechanism for exfoliation, due to the minimal amount of required peeling force and from an analysis of the potential mean force (PMF). The exfoliation analysis is also complemented by an analysis of the fluid structure (density profiles and angle distributions) of each IL on the Bi_2Te_3 surface. Overall, the [C_4mim^+] [Tf_2N^-] solvent provides the lowest barrier for exfoliation, and its effectiveness can be traced to its unique cation and anion interactions with the surface of the Bi_2Te_3 layers.

References

(1) Takashiri, M.; Kai, S.; Wada, K.; Takasugi, S.; Tomita, K. Role of stirring assist during solvothermal synthesis for preparing single-crystal bismuth telluride hexagonal nanoplates. *Materials Chemistry and Physics* **2016**, *173*, 213-218.

(2) Liang, Y.; Wang, W.; Zeng, B.; Zhang, G.; He, Q.; Fu, J. Influence of NaOH on the formation and morphology of Bi2Te3 nanostructures in a solvothermal process: From hexagonal nanoplates to nanorings. *Materials Chemistry and Physics* **2011**, *129*, 90-98.

(3) Li, Z.; Teng, R.; Zheng, S.; Zhang, Y.; Huang, T.; Lu, G. Single-crystalline Bi2Te3 nanosheets with uniform morphology via a simple, efficient, and high-yield microwave-assisted synthesis. *Journal of Crystal Growth* **2014**, *406*, 104-110.

(4) Guo, L.; Ivey, B. C.; Aglan, A.; Tang, C.; Song, J.; Turner, C. H.; Frazier, R. M.; Gupta, A.; Wang, H. T. Vapor Phase Growth of Bismuth Telluride Nanoplatelets on Flexible Polyimide Films. *ECS Solid State Letters* **2012**, *2*, P19-P21.

(5) Kamath, G.; Baker, G. A. In silico free energy predictions for ionic liquid-assisted exfoliation of a graphene bilayer into individual graphene nanosheets. *Phys Chem Chem Phys* **2012**, *14*, 7929-7933.

(6) Ludwig, T.; Guo, L.; McCrary, P.; Zhang, Z.; Gordon, H.; Quan, H.; Stanton, M.; Frazier, R. M.; Rogers, R. D.; Wang, H. T.; Turner, C. H. Mechanism of bismuth telluride exfoliation in an ionic liquid solvent. *Langmuir* **2015**, *31*, 3644-3652.

(7) Lopes, J. N. C.; Deschamps, J.; Padua, A. A. H. Modeling ionic liquids using a systematic all-atom force field. *Journal of Physical Chemistry B* **2004**, *108*, 2038-2047.

(8) Kamath, G.; Baker, G. A. Are ionic liquids suitable media for boron nitride exfoliation and dispersion? Insight via molecular dynamics. *RSC Advances* **2013**, *3*, 8197.

(9) Morishita, T.; Okamoto, H.; Katagiri, Y.; Matsushita, M.; Fukumori, K. A high-yield ionic liquid-promoted synthesis of boron nitride nanosheets by direct exfoliation. *Chem Commun* **2015**, *51*, 12068-12071.

459237 Tuning the Saw-Tooth Tensile Response and Toughness of Multiblock Copolymer Diamond Networks

Abstract:

Liquid crystalline elastomers consisting of block copolymer chains end-linked into a diamond network have recently been shown to have a saw-tooth tensile response to uniaxial deformation, which can provide the basis to design super-tough, highly stretchable materials. When deformed, these systems are forced to increase the interfacial area between blocks, increasing the stress which is only released when chain hairpins unfold and new segregated domains form, leading to the saw-tooth tensile response. We use Molecular Dynamics to study different chain designs to predetermine the number of "teeth" and the strain-orientation dependence of the tensile response. Two chain designs having distinct tensile behaviors have been identified. First, in a ABA (15%-70%-15%) chain with lower backbone stiffness, the stress response is found to be independent of the extensional direction by virtue of morphological rearrangements that occur at low strains, before the saw-tooth behavior ensues. Second, a pentablock chain ABACA (20%-20%-20%-20%-20%) is found to exhibit a sawtooth response with twice as many "teeth" for a given strain. This pentablock architecture significantly decreases the hairpin length, which has the effect of creating two separate sets of unfolding chains in the network, and decreasing the relaxation of the network when new domains form. Unlike other architectures, the number of hairpins in these pentablocks does not monotonically decrease upon deformation, but can plateau or even increase due to hairpin regeneration events. We also explore the realistic synthesis of these diamond networks, with the goal of enhancing the saw-tooth behavior, network extensibility and cross-link mobility.

467778 Impact of Surfaces Charge Distribution on the Mechanism of Heterogeneous Ice Nucleation

Abstract:

Understanding the role played by solid surfaces in ice nucleation is a significant step toward designing surfaces that can control ice nucleation. This is relevant to various fields such as atmospheric chemistry, transportation and food industry. Our overall goal is to explain the mechanisms through which surfaces affect ice nucleation and growth. Such understanding can guide the development of predictive models correlating surface properties, and interfacial water structure and dynamics to the probability of ice nucleation on that surface. To this end, in the present study we focus on the effect of surface charge distribution on the ice nucleating ability of silver iodide (AgI) surfaces. AgI is one of the most potent ice nucleating agents and has been used for cloud seeding. Our simulations indicate that the ice nucleation rate is significantly affected by the charge distribution of the surface. We can explain these effects by considering the effects of surface charges on the interfacial water orientations near the different surfaces. Our results indicate that in addition to lattice spacing, other factors such as charge distribution play a critical role in determining the ice nucleation efficiency of a surface. We further extend this understanding to elucidate the ice nucleating efficiency of mineral surfaces such as kaolinite.

467914 The Role of Dispersion Interactions, Electrostatics, and Entropy in the Interfacial Behavior of MoS_2

Abstract:

The synthesis and applications of two-dimensional transition metal dichalcogenides such as molybdenum disulfide (MoS_2) in membranes, sensors, and microfluidic devices, involves an inevitable contact between these materials and various liquids. The existence of partially ionic bonds in MoS_2, as opposed to covalent bonds in graphene, suggests a plausible role for polar (electrostatic) interactions in determining the interfacial behavior on two-dimensional MoS_2 surfaces. Surprisingly, not only the equilibrium contact angle, which depends solely on the total interaction energy between the surface and the liquid, but also the friction coefficient and slip length, which depend on the spatial variations in the interaction energy, are independent of the inherent polarization in MoS_2. While the former is found to result from the negligible electrostatic interactions between MoS_2 and liquids, the latter results from the tri-layered sandwich structure of the MoS_2 monolayer. Specifically, the tri-layered sandwich structure results in spatial variations in the dispersion interactions that dominate over those in the electrostatic interactions. Furthermore, the wettability of MoS_2 is found to be controlled by a delicate balance between dispersion interactions and entropy. Our findings reveal that in spite of the existence of heteropolar bonds in MoS_2 compared to the existence of homopolar bonds in graphene, there is an unexpected similarity between the interfacial behaviors of these two surfaces. This finding, in turn, elucidates the similar physiosorption behavior of MoS_2 and graphene, including the use of similar liquids to exfoliate these 2D materials.

470710 Interfacial Properties of Model Systems Using Molecular Dynamics Based Framework

Abstract:

The behavior of fluids at solid surfaces plays a significant role within numerous natural phenomena and industrial applications. To this end, understanding and predicting the interfacial properties of these systems is crucial for the development of emerging technologies. Within our group, we have developed an interface potential based approach for determining interfacial properties. In previous work, we performed simulations within a grand canonical ensemble and calculated the interface potential using transition matrix Monte Carlo methods. However, there are limitations to working within this ensemble, particularly when applying the technique to complex molecules and working at relatively low temperatures. Recently, we developed a means for calculating the interface potential using an Isothermal Isobaric ensemble (NPT). In this presentation, we discuss how to implement this approach within a molecular dynamics based framework. The general approach is employed within a "spreading" and "drying" framework to calculate wetting properties, such as the spreading/drying coefficient, interfacial tension and contact angle. We used umbrella sampling to sample states along the order parameter path (defined by volume). We show ways of processing the acquired data using pyMBAR software and a force integration technique to construct an interface potential. Molecular dynamics simulations have been performed using the GROMACS and LAMMPS packages. Results are presented for a model system consisting of fluid-fluid interactions described by Lennard-Jones potential on a structureless 9-3 surface (GROMACS) and an atomistically detailed substrate (LAMMPS). The results obtained using molecular dynamics are compared with those obtained from NPT Monte Carlo simulation.

472217 Estimating Interfacial and Line Tensions Between Water, Vapor and Solid Substrates

Abstract:

An accurate estimation of the interfacial properties of water adjacent to solid substrates is crucial in characterizing wetting in important contexts such as superhydrophobicity, electrowetting, and heterogeneous nucleation. When a liquid droplet makes contact with a solid surface, its characteristic shape and the macroscopic properties such as the contact angle (CA) and the contact angle hysteresis (CAH) depend on the solid-liquid and liquid-vapor interfacial tensions and the three-phase line-tension. The conventional methods of measuring CA, which involve simulation of cylindrical or spherical droplets to obtain equilibrium density profiles, are often sensitive to the droplet size and somewhat arbitrary choice of the location of three-phase contact line. The interface potential method provides an elegant way to overcome these challenges, and accurately determine CA. Here we illustrate the use of the INdirect Umbrella Sampling (INDUS) method to facilitate calculation of the interface potential using molecular dynamics simulations. We use INDUS to characterize the surface tension and CA of water on a range of substrates with a range of attractions. While the estimation of CA through various methods results in relatively small discrepancies, there is an astounding lack of consensus in estimates of the three-phase contact line tension. While most simulations and experiments predict a negative value of line tension of water on hydrophobic substrates, estimates can vary by several orders of magnitude. Furthermore, thermodynamic stability requires that analogous to surface tension, line tension must always be positive. Moreover, there is a lack of molecular level understanding of line-tension, and consequently of the phenomena that depend on it such as contact line pinning, CAH and even heterogeneous nucleation. Here, we suggest ways to characterize and interpret this elusive three phase contact line-tension of water with various substrates.

467824 Molecular Dynamics Study on the Effects of Nanoscale Roughness on the Wear of Alkylsilane Monolayers

Abstract:

Surface functionalization by silane-based monolayers represents a promising lubrication scheme for micro- and nanoelectromechanical systems (MEMS/NEMS) in which high surface area-to-volume ratios can readily induce wear of the high-energy silica surfaces. However, rapid degradation of monolayer coatings has hindered the effectiveness of these materials to date. Studies of the durability of monolayers with varying interfacial binding strengths suggest a prominent mechanism of wear may be through mechanically induced scission of surface bonds [1,2]. As a result, monolayer durability has been shown to be highly sensitive to surface features, such as roughness and crystallinity. To ascertain the feasibility of monolayer coatings for MEMS/NEMS lubrication, a comprehensive understanding of the wear process, including conditions that promote degradation and how these might be overcome, is imperative. Molecular dynamics simulations provide a means to analyze these systems at the atomic-scale. In previous work, molecular dynamics simulations of alkylsilane monolayers were used to show that even at the atomic level, chains located at "peaks" on the surface (positive deviations from the surface mean) are preferentially removed under shear [3]. Nanoscale roughness, emerging from surface asperities, is likely to have an even greater effect on monolayer stability, as a result of the high local pressures arising at asperity contacts. This is exemplified through AFM measurements, which mimic a single-asperity contact, where irreversible monolayer damage has been shown to occur beyond a critical normal load as a result of the breaking of surface bonds [4].

In this work, we examine the wear of alkylsilane monolayers at both single- and dual-asperity contacts. A pseudo-reactive potential is developed to model interfacial bonding, and bond strengths are tuned to investigate the necessary strength required to sustain the high local pressures under these conditions. The influence of monolayer structure (e.g., composition and chain length) on monolayer durability is examined. Our results reveal greater durability as monolayer chain length is increased, agreeing with experiment, but also show that these conditions exhibit a higher friction coefficient, as a result of increased molecular plowing. The effects of surface deformation under single- and dual-asperity contact are also explored.

[1] B. D. Booth, S. G. Vilt, C. McCabe, and G. K. Jennings, "Tribology of Monolayer Films: Comparison between n-Alkanethiols on Gold and n-Alkyl Trichlorosilanes on Silicon.," Langmuir, vol. 25, no. 17, pp. 9995–10001, Sep. 2009.

[2] S. P. Pujari and H. Zuilhof, "Highly wear-resistant ultra-thin perfluorinated organic monolayers on silicon(111) surfaces," Appl. Surf. Sci., vol. 287, no. December, pp. 159–164, Dec. 2013.

[3] A. Z. Summers, C. R. Iacovella, M. R. Billingsley, S. T. Arnold, P. T. Cummings, and C. McCabe, "Influence of Surface Morphology on the Shear-Induced Wear of Alkylsilane Monolayers: Molecular Dynamics Study," Langmuir, 2016.

[4] B. Bhushan, T. Kasai, G. Kulik, L. Barbieri, and P. Hoffmann, "AFM study of perfluoroalkylsilane and alkylsilane self-assembled monolayers for anti-stiction in MEMS/NEMS," Ultramicroscopy, vol. 105, no. 1–4, pp. 176–188, Nov. 2005.

471472 Reaxff Reactive Force Field Study of Cesium Adsorption in Micaceous Clay Minerals

Abstract:

Among the radionuclides released into the environment during the Fukushima Dai-ichi Nuclear Power Plant (FDNPP) accident, ^{137}Cs is the most significant long-term contributor to environmental contamination due to its high release rate (~ 1.3×10^{16} Bq) and longer half-life (30.1 years). High affinity of Cs^{+} absorption in micaceous clay minerals through ion-exchange with native interlayer cations is well known. Knowledge of Cs binding energetics to various clay mineral surfaces is critical to developing effective decontamination strategies. We perform metadynamics simulations to study the adsorption of Cs on the (001), (010) and (110) surfaces of kaolinite, vermiculite and muscovite clay minerals as a function of temperature in pure and marine water under acidic/basic conditions, using a recently developed ReaxFF reactive force field. Our results demonstrate that the adsorption of Cs on various clay mineral surfaces depends critically on the surface corrugation, coordination of the surface atoms and their mutual separations and arrangement in space. The free-energy profiles from the metadynamics simulations were used to calculate the solid-liquid partition coefficient Kd for Cs on the various clay mineral surfaces. Cs was found to adsorb the strongest on the (110) surfaces, followed by the (001) and (010) clay mineral surfaces. Marine solution at pH 7 is the most energetically favorable environment to remove Cs from the studied clay surfaces. Weathered micaceous clay minerals are known to play a crucial role on adsorption and retention of cesium. They have special adsorption sites, called frayed edge sites (FESs), which adsorb cesium selectively and irreversibly. We investigate Cs adsorption at FESs in muscovite surface at various interlayer separations using metadynamics simulations to obtain the Cs/K exchange energies. We observe that Cs adsorption at FESs is significantly more energetically favorable (up to a factor of 10) compared to adsorption on the unweathered clay mineral surface. A critical interlayer separation is necessary for Cs uptake in clay minerals. At lower interlayer separation Cs/K exchange is unfavorable and at wider interlayer separations the FESs have adsorption sites suitable for Cs ions that have a larger radius than the K ions. We reveal the

adsorption mechanism of Cs at FESs and show that multiple distinct intermediate states can occur during Cs uptake in clay minerals.

471511 DFT Calculations of Multi-Coverage and Phase Adlayer Structure of Tryptophan on Cu(111)

Abstract:

In this work, we present results of DFT calculations, in conjunction with recent experiments via STM, to characterize the structure of low to moderate to high coverage adlayers

of the amino acid tryptophan on Cu(111) surfaces. This work characterizes the differences in structure between chiral enantiomeric

mixture dosings of the tryptophan on Cu(111). Because tryptophan has a large steric R-group, the 'universally' accepted assumption

of tridentate bonding surface bonding motifs may not be valid in this system; our results help elucidate the truth of this assumption for this (and possibly other related) systems. Current results indicate that long held assumptions about the tri-dentate chiral bonding motif footprint for amino acids on copper surfaces may be situationally dependent.

468541 A Theoretical Study of Bulk and Inhomogeneous Hydration of Primary Amines

Abstract:

A Theoretical Study of Bulk and Inhomogeneous Hydration of Primary Amines

Shijie Sheng, Jia Fu and Jianzhong Wu*

Department of Chemical and Environmental Engineering,

University of California, Riverside, CA 92521

Abstract

Amine functionalized surfaces are relevant to various industrial applications, for example, dye decomposition, carbon dioxide fixation and heavy metal enrichment from waste water. The surface effect on amine hydration plays a pivotal role in particular for applications with surface reactions. This work aims to unravel the change in the hydration structure of primary amines due to the presence of a silica surface at different solution conditions. To calibrate our theoretical predictions based on the molecular density functional theory, all-atom molecular dynamic (MD) simulations are first applied to a silica slab with amine molecules tethered onto a silica surface at representative pH values. In comparison to amine solvation in the bulk phase, both theoretical and simulation results indicate that the silica surface alters the hydration structure sensitive to the pH and the chain length of the primary amine. The effects of silica surface on the hydration-bonding dynamics are also studied.

---

* Email: jwu@engr.ucr.edu

475314 Designing Functional Materials within the Materials Project

Abstract:

The Materials Project (www.materialsproject.org) â€“ part of the broader Materials Genome Initiative - is an effort to compute the properties of all known inorganic materials and beyond, and offer that data to the community together with online analysis and design algorithms. The current release contains data derived from density functional theory (DFT) calculations for over 66,000 materials, each with searchable associated properties such as relaxed structure, electronic state, energy storage capability, aqueous and solid stability, and more. Furthermore, software algorithms are offered by the Materials Project and can be used by researchers for materials informatics, including both interactive web-based tools like the Phase Diagram App and the Pourbaix App, as well as open-source codebases and data access tools such as the pymatgen materials analysis library, FireWorks workflow software, and Materials API.Â Today â€“ 4 years after launch, the Materials Project is driving materials innovation in broad chemical and structure spaces, for applications as varied as energy storage, energy production, thermoelectricity, transparent conductors, materials synthesis conditions etc. This talk will highlight the development of the Project, its growth attracting more than 20,000 users and a few of the many projects that have been successfully concluded or are being pursued using the capabilities and materials understanding that has emerged from our approach of data informed materials design. Such projects are i) stability of inorganic materials in aqueous electrolytes for battery, fuel cell and catalysis applications, and design principles for ii) novel multivalent intercalation cathode discovery and iii) electrolytes.

473787 Data Analytics and Machine Learning in Nanomaterials Discovery

Abstract:

The recent explosion of "big data" has positioned data analytics and machine learning techniques at the centre of social, marketing, medicine and manufacturing research. In nanomaterials discovery, data analytics tools are also playing a fundamental role to successfully tackle the exponential increase in size and complexity of functional nanomaterials. In this work, the intrinsic relationships between nanoscale features and functional properties of nanomaterials will be discussed for nanodiamonds[1], graphene nanoflakes[1-3] and nanoporous metal-organic frameworks (MOFs)[4-6]. Multivariate statistics are applied to identify the most relevant individual features in conjunction with the most representative nanostructures in nanoparticle ensembles and structural libraries. Meanwhile, machine learning techniques, from simple decision tree predictors to more complex suppert vector machines and neural networks, are utilized to predict the electronic properties of graphene nanoflakes and to identify "high-performing" nanoporous MOFs for carbon capture and storage applications. It will be demonstrated that simple predictors provide useful insights into the structure-property relations paradigm at the nanoscale, whilst more complex machine learning models are efficient tools to rapidly discriminate among potential candidate materials at a fraction of the traditional computational cost.

References

1. Fernandez, M. & Barnard, A. S. ACS Nano 9, 11980–11992 (2015).

2. Fernandez, M., Shi, H. & Barnard, A. S. J. Chem. Inf. Model. 55, 2500–2506 (2015).

3. Fernandez, M., Shi, H. & Barnard, A. S. Carbon 103, 142–150 (2016).

4. Fernandez, M., Boyd, P. G., Daff, T. D., Aghaji, M. Z. & Woo, T. K. J. Phys. Chem. Lett. 5, 3056–3060 (2014).

5. Fernandez, M. & Barnard, A. S. ACS Comb. Sci. (2016). doi:10.1021/acscombsci.5b00188

6. Aghaji, M. Z., Fernandez, M., Boyd, P. G., Daff, T. D. & Tom, K. Eur. J. Inorg. Chem. (2016) in press.

472740 Describing the Diverse Geometries of Gold from Nanoclusters to Bulk a First-Principles Based Hybrid Bond Order Potential

Abstract:

Gold nano structures are of intense interest for catalysis, sensing, plasmonic, and other applications, and gold nanoclusters (<200 atoms), in particular, exhibit unusual chemistry in the nano-scale. In gold nanoclusters, a fundamental understanding of the atomic structure, size-dependent dimensionality, surface chemistry, and atomic-scale dynamics over timescales spanning tens of nanoseconds holds the key to unraveling structure-property relationships. Gold nanoclusters undergo a fascinating size dependent transition in cluster dimensionality and display a wide range of structural motifs (e.g., planes, hollow cages, tubes, space-filled compact forms). The origin of such transitions has been largely unknown; previous density functional theory (DFT) and spectroscopy studies have speculated that while bond directionality effects predominantly govern structure/chemistry in small Au clusters, long-range dispersive forces probably become important in large clusters, surfaces and bulk condensed phases. A major roadblock has been our inability to perform long time molecular dynamics simulations, and global optimization of structures due to lack of accurate empirical force fields.

In this talk, we introduce a unified *ab initio*-based hybrid bond order potential (HyBOP), which captures the competition between short-range interactions (bond directionality) and long-range dispersions to accurately predict the size dependent diverse geometries of gold clusters, surfaces and bulk. Our work involves an important methodological development: we employed genetic algorithms to train our HyBOP potential against a large DFT dataset comprising (a) energies of a thousand 13-atom Au clusters, (b) equation of state for various bulk polymorphs of gold, and (c) low-index surfaces of bulk face-centered gold. Our newly developed HyBOP accurately predicts (a) evolution of various structural motifs in Au clusters with increasing size, (b) critical size for transition from planar to globular structures (*i.e.*, dimensionality changes), (c) global energy minimum configurations at various cluster sizes, and (d) thermodynamics, structure, elastic properties, and energetic ordering of bulk condensed phases as well as surfaces, in

excellent agreement with DFT calculations and spectroscopic experiments. As a representative test, using long time HyBOP based MD simulations, we identified the atomic scale mechanisms governing the coalescence of icosahedral and planar clusters into pyramids.

469402 Topological Data Analysis of Nanoporous Materials Genome Using Pore-Geometry Recognition Technique

Abstract:

In most applications of nanoporous materials the pore structure is as important as the chemical composition as a determinant of performance. For example, one can alter performance in applications like carbon capture or methane storage by orders of magnitude by only modifying the pore structure. For these applications it is therefore important to identify the optimal pore geometry and use this information to find similar materials. However, the mathematical language and tools to identify materials with similar pore structures, but different composition, has been lacking. Here we develop a pore recognition approach to quantify similarity of pore structures and classify them using topological data analysis. Our approach allows us to identify materials with similar pore geometries, and to screen for materials that are similar to given top-performing structures. Using methane storage and carbon capture as a case study, we also show that materials can be divided into topologically distinct classes — and that each class requires different optimisation strategies. Inthis work we have focused on pore space, but our topological approach can be generalised to quantify similarity of any geometric object, which, given the many different Materials Genomics initiatives, opens many interesting avenues for big-data science.

468015 Unraveling the Role of Pore Topology and Chemical Functionality on the Carbon Capture Performance of Metal-Organic Frameworks

Abstract:

Metal-organic frameworks (MOFs) are novel porous materials obtained via assembly of inorganic and organic molecular building blocks into crystalline networks. Thousands of MOFs have been synthesized to date, based on different combination of building blocks. The evident structural and chemical versatility of MOFs has made them promising materials for applications such as CO_2 capture in fossil-fuel-based power stations. The challenge, however, is to efficiently identify high-performing materials within the vast MOF "structure space." Using molecular modeling, here we selectively explore this space aiming to elucidate the role of pore topology and chemical functionality in the CO_2 capture performance of MOFs to establish clear design rules that can guide the discovery and synthesis of optimal MOFs for this application.

The chemical functionality of MOFs has long been deemed key to enhance CO_2 adsorption in MOFs. However, since the incorporation of functional group also alters MOF textural properties, conflicting information on the positive or negative impact of a chemical moiety on CO_2 adsorption often arises. In addition, elucidating optimal functionalities experimentally has been difficult, because comprehensive sets of MOFs with identical structures but different functionality are not readily synthesizable or otherwise equally stable upon activation and testing. Similar issues are encountered in attempting to elucidate optimal pore topologies for CO_2 adsorption. The above roadblocks can be circumvented using molecular simulation to evaluate a suitable, comprehensive, diverse set of "perfect MOFs" but the challenge is to efficiently create and evaluate such predictably large set in the computer.

In this work, we created *in silico* 426 MOFs carrying different simple chemical functionalities, based on 28 "parent" MOFs of 16 different topologies, using our recently developed automated MOF construction algorithm. Since CO_2 adsorption is likely to be highly sensitive to Coulombic interactions, we explored different strategies to assign partial charges to MOF atoms, including a new building block-based strategy that we introduce in this work to be compatible with our MOF construction algorithm. Based on our simulations, we provide insights into how and under what conditions different chemical functionalities and topologies are likely to enhance CO_2 adsorption. For instance, from our data analysis we find that at pressures relevant to post-combustion CO_2 capture, alcohol and thiol functionalities tend to outperform other moieties often suggested to enhance CO_2 adsorption such as amino functionalities. Additionally, we present potentially attractive MOF synthesis targets based on CO_2 adsorption properties predicted during our computational screening.

460977 An Automated Approach for Developing Graph-Theoretical Cluster Expansions of the Total Energy of Adsorbed Layers

Abstract:

An Automated Approach for Developing Graph-Theoretical Cluster Expansions of the Total Energy of Adsorbate Layers

E. Vignola[1,2], S. N. Steinmann[2], M. Stamatakis,[3] P. Sautet[2,4]

[1]TOTAL Petrochemicals, Route de la Chimie,76700 Gonfreville-l'Orcher, France; [2] Univ Lyon, Ens de Lyon, CNRS, Université Lyon 1, Laboratoire de Chimie UMR 5182, F-69342, Lyon, France; [3] Department of Chemical Engineering, University College of London, Torrington Place, London WC1E7JE, United Kingdom; [4] Department of Chemical and Biomolecular Engineering, University Of California Los Angeles, CA, USA

The accurate description of the total energy of adsorbate layers is crucial for the understanding of chemistry at interfaces. For catalysis applications in particular, adsorbate-adsorbate lateral interactions have been shown to significantly affect activation energies of reactions, thereby shaping experimentally observed trends. [1,2] Modelling the interactions of atomic adsorbates has traditionally been achieved using effective Ising-type Hamiltonians [3], whereby a set of spin-like values is attributed to the layer's lattice points describing the occupancy of the corresponding catalytic sites (vacant occupied by a species). Pairwise additive adsorbate-adsorbate lateral interactions in this model are captured by appropriate coupling constants.

Such a Hamiltonian is however limited, as it cannot account for adsorbates that bind to more than one sites (bi-dentate or even multi-dentate species), and it cannot capture many-body contributions to the total energy (3-body interactions

– triplets). To overcome these limitations one has to adopt a cluster expansion Hamiltonian formalism [4], which has recently been implemented in a graph-theoretical scheme [5,6] to enable the representation of multi-dentate species. Automating the development of such cluster expansion Hamiltonians for catalytic systems is challenging. Existing approaches for such automation [7] can only tackle mono-dentate adsorbates and cannot account for the various binding modes that molecules exhibit on solid surfaces.

The current work develops a scheme for automating the development of cluster expansions applicable to molecular species on catalytic surfaces. The scheme has been implemented in a FORTRAN 95 program compatible with the graph-theoretical kinetic Monte Carlo code, *Zacros* [8].

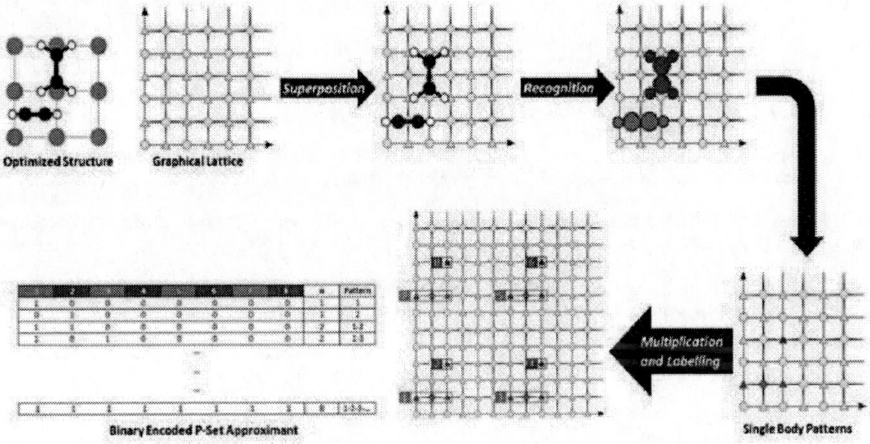

Figure 1. Pattern Recognition and Encoding Scheme

References

[1] Stamatakis, M.; Piccinin, S. *ACS Catalysis* **2016**, *6*, 2105

[2] Frey, K.; Schmidt, D. J.; Wolverton, C.; Schneider, W. F. *Catal. Sci. Technol,* **2014**, *4*, 4356

[3] Mussardo, G. *Statistical Field Theory. An Introduction to Exactly Solved Models in Statistical Physics,* Oxford University Press, New York, 2010.

[4] Sanchez, J. M.; Ducastelle, F.; Gratias, D. *Phys. A Stat. Mech. its Appl.* **1984,** *128 (1-2)*, 334.

[5] Stamatakis, M.; Vlachos, D. G. *J. Chem. Phys.* **2011**, 134, 214115

[6] Nielsen, J.; d'Avezac, M.; Hetherington, J.; Stamatakis, M. J. Chem. Phys. **2013**, *139 (22)*, 224706

[7] van de Walle, A.; Ceder, G. *J. Phase Equib.* **2002**, 23, 348.

[8] Stamatakis M. *Zacros: Advanced Lattice-KMC simulation Made Easy*, http://www.zacros.org/, 2013.

453762 Property Prediction of Crystalline Solids from Composition and Crystal Structure

Abstract:

Property Prediction of Crystalline Solids from Composition and Crystal Structure

Bruno A. Calfa[1] and John R. Kitchin[2]

Predicting properties of compounds has received considerable attention across different disciplines and has seen applications in diverse areas. The approach proposed in this paper is influenced by the advances in Group Contribution Methods (GCMs).

A key characteristic of GCMs in general is that the property descriptors (i.e., predictors) are explicitly represented by the molecular structure (and chemical composition) of a compound. One can pose an optimal *inverse* (design) problem, which attempts to obtain the original molecule (structure and composition) given a target property value. This is the main goal of the research area called Computer-Aided Molecular/Mixture Design (CAMD) as exemplified by several works in the literature (Eljack et al., 2007; Samudra and Sahinidis, 2013).

In the crystalline solids literature, several research groups have proposed statistical (machine learning) approaches for property prediction. Saad et al. (2012) investigated both unsupervised and supervised machine learning techniques to predict structure and properties of crystals with chemical formula AB. Ma et al. (2015) developed a machine-learning-augmented model based on artificial neural networks (ANNs) that captures nonlinear adsorbate-substrate interactions.

We propose using kernel regression (Li and Racine, 2007) as a data-driven and rigorous nonparametric statistical technique to predict properties of atomic crystals. A key feature of the proposed approach is the possibility of treating

predictors not only as continuous, but also as categorical data. The latter specifically allows the predictive model to capture the discrete nature of crystals with regards to composition (number of atoms in the chemical formula) and spatial configuration (finite number of crystallographic space groups). Another important aspect of using kernel regression is the direct access to its explicit mathematical form, which can be directly embedded in optimal inverse problems to design new crystalline materials with given target properties. The property prediction approach is illustrated by training models to predict electronic properties of 746 binary metal oxides and elastic properties of 1,173 crystals. As a first approach to solving the inverse problem, we describe an exhaustive enumeration algorithm (Calfa and Kitchin, 2016).

References

Calfa, B. A.; and Kitchin, J. R. 2016. Property Prediction of Crystalline Solids from Composition and Crystal Structure. AIChE Journal. In press. DOI: 10.1002/aic.15251.

Eljack, F. T.; Eden, M. R.; Kazantzi, V.; Qin, X.; and El-Halwagi, M. M. 2007. Simultaneous Process and Molecular DesignÑA Property Based Approach. AIChE Journal. 53(5):1232Ð1239.

Li, Q., and Racine, J. S. 2007. Nonparametric Econometrics: Theory and Practice. Themes in Modern Econometrics. Princeton University Press. New Jersey, NJ. USA.

Ma, X.; Li, Z.; Achenie, L. E. K.; and Xin, H. 2015. Machine-Learning-Augmented Chemisorption Model for CO2 Electroreduction Catalyst Screening. Journal of Physi- cal Chemistry Letters. 6(18):3528Ð3533.

Saad, Y.; Gao, D.; Ngo, T.; Bobbitt, S.; Chelikowsky, J. R.; and Andreoni, W. 2012. Data Mining for Materials: Computational Experiments with AB Compounds. Physical Review B. 85(10):104104.

Samudra, A. P., and Sahinidis, N. V. 2013. Optimization-Based Framework for Computer- Aided Molecular Design. AIChE Journal. 59(10):3686Ð3701.

---

[1] Department of Chemical and Biological Engineering. University of Wisconsin-Madison. Madison, WI 53706. USA.

[2] Department of Chemical Engineering. Carnegie Mellon University. Pittsburgh, PA 15213. USA.

453137 Local Pattern Discovery for Uncovering Structure-Property Relationships of Materials

Abstract:

Machine learning applications to data from atomistic material simulations typically focus on the inference of a global prediction model for some physical or chemical property of interest, such as activation barriers or binding energies. Due to their global prediction-based objective, these models are not well suited for the discovery of physically interpretable characterizations describing sub-groups of material configurations that share common properties. In contrast to predictive global modeling, local pattern discovery techniques directly aim to uncover specific local properties of the data. Moreover, they describe these properties through models that are built from discrete symbols corresponding to meaningful notions of the discovery domain, e.g., find sub-groups of materials with large band gaps and low formation energies. Therefore, they constitute a complimentary part of the data mining toolbox for the automatic analysis of materials science data repositories, such as for the Novel Materials Discovery Repository. As an exemplary application of local pattern discovery for materials science, we consider the automatic discovery of structure-property relationships of gold clusters (size 7-14). Local patterns between geometrical (e.g., local bond coordination, cluster size, and radius of gyration) and physicochemical (e.g., HOMO-LUMO gap and intramolecular van der Waals energy) properties are found.

464732 Alloy Catalyst Discovery Using Computational Alchemy

Abstract:

There is great interest in finding identifying high-performance catalysts that are economical. Computational quantum chemistry schemes employing Kohn-Sham density functional theory (KS-DFT) can be used to screen catalyst materials on the basis of thermodynamic descriptors (see for example ref. [1]). Though usually considered reliable for descriptor-based analyses, KS-DFT calculations are computationally expensive and intractable for use when screening across the full chemical space of all possible alloy materials. Toward this goal, we employ a model Hamiltonian method, Ôcomputational alchemyÕ [2-4] to approximate KS-DFT energies at a fraction of the computational cost. We will introduce the theory of computational alchemy and how it can be used to facilitate descriptor-based screening on many thousands of alloys structures.

Reference:

1. Greeley, J., Jaramillo, T. F., Bonde, J., Chorkendorff, I. & N¿rskov, J. K. Computational high-throughput screening of electrocatalytic materials for hydrogen evolution. Nat Mater 5, 909Ð913 (2006).

2. Lilienfeld, O. A. von, Lins, R. D. & Rothlisberger, U. Variational Particle Number Approach for Rational Compound Design. Phys. Rev. Lett. 95, 153002 (2005).

3. Lilienfeld, O. A. von & Tuckerman, M. E. Molecular grand-canonical ensemble density functional theory and exploration of chemical space. The Journal of Chemical Physics 125, 154104 (2006).

4. Sheppard, D., Henkelman, G. & Lilienfeld, O. A. von. Alchemical derivatives of reaction energetics. The Journal of Chemical Physics 133, 084104 (2010).

470073 To Address Surface Reaction Network Complexity Using Machine Learning, Scaling Relations, and DFT Calculations

Abstract:

Surface reaction networks involving hydrocarbons exhibit enormous complexity with hundreds of species and thousands of reactions for all but the very simplest of chemistries. Among fields that regularly confront this challenge, such as catalysis, metabolic engineering, and combustion, catalysis is unique in that the most likely reaction pathway varies for each surface and active site and must be re-determined in every instance proving a fundamental bottleneck in new catalyst discovery. We present a framework to efficiently explore a reaction network from scratch using a hierarchy of well-established predictive methods such as group additivity and linear scaling relations as guides to traditional density functional theory (DFT) approaches. Unexplored regions of the reaction network are predicted using machine learning methods, and the mechanism is progressively refined using uncertainty-aware DFT calculations on the parts of the network that are likely rate-limiting. Applying these methods to the reaction of syngas on rhodium (111), we identify the most likely reaction mechanism along with bounds on the likelihood unexplored reaction pathways do not contribute, using DFT on just a fraction of the possible intermediates. Propagating uncertainty throughout this process allows for estimates that the final mechanism is complete given DFT-level uncertainty and measurements on only a subset of the entire network.

467637 Machine-Learning Acceleration of the Exploration of Potential Energy Surface within Amp

Abstract:

Novel nanostructures with enhanced properties are required to be computationally designed in various fields of science and engineering, e.g. nanomaterials with improved load-bearing properties in materials science [1] or catalysts to speed up reaction kinetics in chemistry [2]. In principle, accurate calculations can be carried out based on electronic structure methods. However, with the current status of computational facilities, such accurate calculations are feasible only on relatively small systems within limited time scale. In other words the accuracy might have to be sacrificed in favor of less computational expense as the length and time scales of simulations increase. As the result, emergence of new computational methods seems essential.

Pattern recognition of energetics from an *ab initio* database can help reducing the number of required quantum mechanical calculations, as the result making larger-length or longer-time scale simulations practical. This can be done inside our open-source modular Atomistic Machine-learning Package (Amp) [3], which has been recently developed by the authors of this paper. Amp has two main components, a descriptor for modeling local atomic environments and a regression model to establish the relationship between energies/forces with local atomic environment. Amp is developed compatible with the widely-used open-source Atomic Simulation Environment (ASE), facilitating interactive cooperation between the Amp machine-learning calculator and any *ab initio* calculator within ASE. This will be illustrated here in the context of two well-known problems in molecular sciences, a) determination of the structure of stable nanoclusters, and b) finding the minimum energy pathway of reactions. It will be shown that machine-learning can remarkably improve the exploration of the potential energy surface.

References

[1] Q. Qin et al.; Recoverable plasticity in penta-twinned metallic nanowires governed by dislocation nucleation and retraction, Nature Communication, 6, 2015.

[2] S. Freakley et al.; Palladium-tin catalysts for the direct synthesis of H_2O_2 with high selectivity, Science, 351, 2016.

[3] A. Khorshidi, A. A. Peterson; Amp: A modular approach to machine learning in atomistic simulations, Computer Physics Communications, 2016.

AUTHOR INDEX

AUTHOR INDEX

AUTHOR INDEX

AUTHOR INDEX

AUTHOR INDEX

AIChE
120 Wall Street, FL 23
New York, NY 10005-4020

ISBN 978-1-5108-3431-6

CURRAN ASSOCIATES INC.
proceedings.com

Process Development Division 2019

Held at the 2019 AIChE Spring Meeting and
15th Global Congress on Process Safety

New Orleans, Louisiana, USA
31 March - 4 April 2019

AIChE
The Global Home of Chemical Engineers